T0233971

Lecture Notes in Computer Science   10130

Commenced Publication in 1973
Founding and Former Series Editors:
Gerhard Goos, Juris Hartmanis, and Jan van Leeuwen

More information about this series at http://www.springer.com/series/8637

Abdelkader Hameurlain · Josef Küng
Roland Wagner · Klaus-Dieter Schewe
Karoly Bosa (Eds.)

Transactions on Large-Scale Data- and Knowledge-Centered Systems XXX

Special Issue on Cloud Computing

Springer

Editors-in-Chief

Abdelkader Hameurlain
IRIT, Paul Sabatier University
Toulouse
France

Roland Wagner
FAW, University of Linz
Linz
Austria

Josef Küng
FAW, University of Linz
Linz
Austria

Guest Editors

Klaus-Dieter Schewe
Software Competence Center Hagenberg
Hagenberg
Austria

Karoly Bosa
Software Competence Center Hagenberg
Hagenberg
Austria

ISSN 0302-9743   ISSN 1611-3349   (electronic)
Lecture Notes in Computer Science
ISSN 1869-1994
Transactions on Large-Scale Data- and Knowledge-Centered Systems
ISBN 978-3-662-54053-4   ISBN 978-3-662-54054-1   (eBook)
DOI 10.1007/978-3-662-54054-1

Library of Congress Control Number: 2016958995

Printed on acid-free paper

This Springer imprint is published by Springer Nature
The registered company is Springer-Verlag GmbH Germany
The registered company address is: Heidelberger Platz 3, 14197 Berlin, Germany

Preface

Today, many providers of cloud services (IaaS, SaaS, PaaS, DaaS, etc.) emphasize the many benefits of outsourcing applications into a private or public cloud. For instance, cloud computing reduces spending on technology infrastructure, improves accessibility, etc. Moreover, cloud architectures are resilient in scaling IT resources up and down on-demand and cloud services are able to adapt to dynamically changing business requirements.

In other words, it is suggested that cloud computing represents a mature technology that is ready to be massively used; however, cloud computing still requires extensive research. Many actors of academia and industry are concerned about cloud security, privacy, availability, data protection, interoperability, etc.

The TLDKS special issue on Cloud Computing is devoted to disseminating original research contributions. Topics of interest include, but are not limited to:

- Cloud applications, infrastructures, and platforms
- Service-oriented architectures in cloud computing
- Cloud resource selection, composition, provisioning, and allocation
- Cloud service development process modelling
- Service ontologies and knowledge management
- Personalization of cloud services
- Formal models and methods for cloud computing/services computing
- Security and privacy for the cloud
- Advancing cloud SLA management, business models, and pricing policies
- Semantic interoperability techniques for heterogeneous cloud services
- Multi-clouds/federated clouds
- Innovative cloud-based architectures and techniques for big data analytics
- Client- and user-community-centric cloud computing
- Large-scale sensor systems/large-scale cyber-physical systems
- Mobile clouds
- Location-based services, presence, availability, and locality

Reviewers were asked to provide a clear and justified recommendation, whether to accept a paper without or with minor changes, to request a major revision, or to reject the paper. Authors of the papers for which a major revision was requested in the first round were invited to submit the revised version for a second reviewing round. This volume contains five fully revised selected regular papers.

We would like to express our great thanks to the editorial board and the external reviewers for thoroughly reviewing the submitted papers and ensuring the high quality of this volume. Special thanks go to Gabriela Wagner for her availability and her valuable work in the realization of this TLDKS volume.

August 2016

Klaus-Dieter Schewe
Karoly Bosa

Editorial Board

Contents

Contents

Cloud Computing: Read Before Use

Amol Jaikar[1,2(✉)] and Seo-Young Noh[1,2]

[1] Grid and Supercomputing Department,
Korea University of Science and Technology, Daejeon 305-350, South Korea
[2] National Institute of Supercomputing and Networking,
Korea Institute of Science and Technology Information,
Daejeon 305-806, South Korea
{amol,rsyoung}@kisti.re.kr

Abstract. Cloud computing is evolving as a new paradigm in service computing in order to reduce initial infrastructure investment and maintenance cost. Virtualization technology is used to create virtual infrastructure by sharing the physical resources through virtual machine. By using these virtual machines, cloud computing technology enables the effective usage of resources with economical profit for customers. Because of these advantages, scientific community is also thinking to shift from grid and cluster computing to cloud computing. However, this virtualization technology comes with significant performance penalties. Moreover, scientific jobs are different from commercial workload. In order to understand the reliability and feasibility of cloud computing for scientific workload, we have to understand the technology and its performance. In this work, we have evaluated the scientific jobs as well as standard benchmarks on private and public cloud to understand exact performance penalties involved in adoption of cloud computing. These jobs are categorized into CPU, memory, N/W and I/O intensive. We also analyzed the results and compared the private and public cloud virtual machine's performance by considering execution time as well as price. Results show that the cloud computing technology faces considerable performance overhead because of virtualization technology. Therefore, cloud computing technology needs improvement to execute scientific workload.

Keywords: Performance · Scientific workflow · Cloud computing · Virtualization

1 Introduction

Cloud computing is a model for enabling ubiquitous, convenient, on-demand network access to a shared pool of configurable computing resources (e.g., networks, servers, storage, applications, and services) that can be rapidly provisioned and released with minimal management effort or service provider interaction [1]. It has characteristics like on-demand self-service, broad network access, resource pooling and rapid elasticity. Cloud computing offers Infrastructure, Platform and

© Springer-Verlag GmbH Germany 2016
A. Hameurlain et al. (Eds.): TLDKS XXX, LNCS 10130, pp. 1–22, 2016.
DOI: 10.1007/978-3-662-54054-1_1

Software as a Service to the customers based on Service Level Agreement (SLA). It also has capability to attract users as well as entrepreneur to reduce the initial investment cost by leasing the resources on demand from cloud infrastructure providers [2]. Running multiple underutilized virtual machines on a single physical machine can reduce the power consumption cost of the data center. Cloud computing is an extended version of cluster, grid and utility computing with help of web browser and virtualization technology. Due to these advantages, many research institutes, government organizations and industries are trying to adopt cloud computing to solve their increasing computing and storage problem [3].

Virtualization technology enables the cloud computing technology. It acts like a back-bone of the cloud computing. Due to virtualization, cloud computing gets benefits like flexibility, isolation, security and easy management. Multiple operating systems can run on a single physical machine with help of virtualization technology [4]. However, virtualization comes with some performance overhead [5]. Moreover, there are many virtual machine monitors available in the market and each virtual machine runs on this virtual machine monitor. Therefore, the performance of virtual machine is closely based on virtual machine monitor [6]. There are mainly two types of virtual machine monitors, software and hardware [7]. In order to select best virtual machine monitor for our environment, we have to measure the performance. We have chosen two open source technologies Xen and KVM to test not only scientific workload but also standard benchmarks. Xen and KVM are an example of para-virtualization and full-virtualization respectively.

Scientific workload is different from commercial jobs [8]. Scientific computing requires huge amount of resources to calculate the results within a reasonable time frame. Therefore, scientific workload mainly executes on supercomputers. These supercomputers are very expensive to buy as well as to manage. Many projects tried to adopt technologies like cluster and grid on commodity hardware to reduce the cost of the project [9]. In order to manage these resources, scientist requires computer experts. Therefore, cloud computing gives an alternative to scientist to lease the resources for the project with low cost [10]. However, time frame is also an important factor in scientific workload. In order to understand the performance of scientific workload on private and public cloud, we have executed scientific jobs as well as standard benchmarks and measured the execution time.

Performance and price of computing systems are very close in a relation. Consumer has to pay more for high performance systems. Power consumption of these high performance systems of a data center is equivalent to a small city. These types of data centers in U.S.A. consume around 1.2 % of total energy consumption of whole country [11]. Therefore, most of the researchers have devoted their research towards green data center [12]. Around 45 % of total power consumption belongs to servers in a data center [13]. Moreover, virtualization also increases the execution time. Therefore, the performance of cloud computing needs to be verified by offloading consumer's workload.

Our main contribution towards performance evaluation of private and public clouds is categorized in three sections:

- Investigate the performance of private and public cloud by executing scientific workload
- Test the performance of private and public cloud by executing standard benchmarks
- Lastly, we compare and analyze the performance by considering cost and time

This paper covers background information in Sect. 2 which includes overview of virtualization and cloud technology. It also covers the scientific jobs details. Section 3 gives experimental setup and Sect. 4 shows performance evaluation of not only scientific workload, but also standard benchmarks. Same section compares the results of private and public cloud performance. Discussion on the results is also covered in Sect. 4 itself. Related work is described in Sect. 5. Lastly, we present our conclusion in Sect. 6.

2   Background

In this section, we provide background information to clearly understand the motivation of this research. It includes overview of virtualization, types of cloud and scientific workload.

2.1   HTCondor

HTCondor is an open source high throughput computing framework for non-interactive parallel jobs which is developed by the University of Wisconsin, Madison [14]. We have used HTCondor batch system framework to test the scientific jobs. Figure 1 shows the master-slave architecture of HTCondor. User submits non-interactive jobs to the master node. Master node runs scheduler, negotiator, collector and queue daemon. Scheduler daemon schedules user's jobs to the work node. This daemon represents resources requests to HTCondor pool. Collector daemon is responsible for gathering all the information of resources for HTCondor pool. Negotiator daemon negotiates between available resources and HTCondor batch system. It can also prioritize the user's request. HTCondor batch system also has a queue, in which jobs can reside until the physical or virtual computing resources are available.

HTCondor batch system has feature like DAGMan (Directed Acyclic Graph Manager). DAGMan can be used to design the flow of jobs. Sometimes one or more jobs depend on another jobs. In order to handle such situation, DAGMan can specify the execution flow such that the dependencies between jobs can be solved. HTCondor-G has capability to communicate with grids or clouds to distribute the jobs. Even, it can communicate with public clouds. University of Liverpool demonstrated the effective uses of HTCondor batch system with power saving [15]. HTCondor batch system has a responsibility to manage the jobs seamlessly.

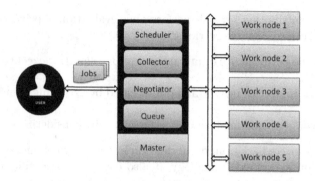

Fig. 1. HTCondor architecture

2.2   Virtualization

Virtualization technology is used to share the physical resources among the virtual machines, which are running on virtual machine monitor or hypervisor. It uses time-sharing technique to run different operating systems on a single physical machine through virtual machine. Virtual machine has its own operating system known as guest operating system. Virtualization gives more benefits to cloud computing like flexibility and isolation. This technology is composed of hardware and/or software component. It creates virtual hardware environment to run guest operating system. Virtual hardware environment consists of virtual CPU, virtual RAM and virtual network interface card. Virtual machine monitor or hypervisor is a middle-ware between hardware and guest operating system.

Virtualization started in 1960 to serve mainframe computers. Later, it has been used for testing environments of operating system research. Mainly two types of virtualization are present: hardware and desktop virtualization. In hardware virtualization, guest virtual machine is totally isolated from host machine. It creates the environment such that the guest operating system is unable to know that it is running on software layer. Desktop virtualization separates logical desktops from physical machine. Hardware virtualization is again divided into three subtypes: full-virtualization, para-virtualization and operating system (OS)-level virtualization.

Full-Virtualization. KVM (Kernel-based Virtual Machine) is an example of full-virtualization with the help of hardware extensions like Intel VT or AMD-V [16]. Figure 2 shows the architecture of KVM. It is originally designed for x86 processors. KVM uses QEMU (Quick Emulator) to virtualize hardware. Full memory virtualization is allowing guest to securely access a physical device [17].

Full virtualization works on dynamic binary translation. A dynamic binary translator is used to translate the privileged operations into unprivileged instructions [18]. These translated unprivileged instructions can run on CPU and get the similar output. Using hardware support like Intel VT and AMD-V, virtual

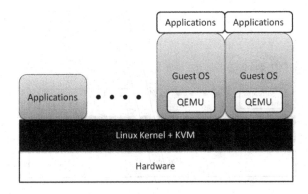

Fig. 2. KVM architecture

machine monitor runs in a new privileged frame. Therefore, Intel VT and AMD-V solve the CPU virtualization problem with considerable overhead. We have used KVM in our experiments to test the scientific jobs as well as standard benchmarks.

Para-Virtualization. In this technique, it creates the software environment which is similar but not identical to the hardware. This technique tries to reduce the performance overhead, which is present in full-virtualization. Xen is an example of para-virtualization [19] which is developed by University of Cambridge. Figure 3 shows the arechitecture of Xen. It is a micro-kernel design which consists of dom0 and domU. Only dom0 have direct access of hardware. From dom0, the hypervisor manages the unprivileged virtual machines (domU) [20]. In para-virtualization technique, the guest operating system needs to be modified to run on para-virtualized hypervisor. However, recently Xen started supporting unmodified operating systems.

Para-virtualization provides hypercalls for guest virtual machine's operating system to communicate with virtual machine monitor. Due to less overhead, it has been stated that para-virtualization is more effective and faster than full-virtualization. Because of modification requirement of operating system, para-virtualization has limited usage in the industry. In our experiment, we have used unmodified operating system for virtual machine.

OS-Level Virtualization. OpenVZ is an example of OS-level virtualization [21]. OpenVZ is container-based virtualization for Linux. In this technique, the kernel of the host operating system grants to create isolated user space known as containers. Each container has root access, IP address, memory and many more, same like normal server. OpenVZ is limited to Linux operating system for containers. Memory allocation is soft, which means that memory not used by one container can be used by other. It is very fast and efficient.

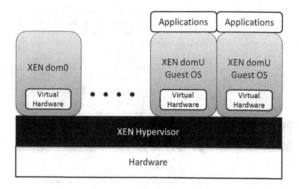

Fig. 3. Xen architecture

Operating system level virtualization partitions the physical resources at operating system level. Therefore, all containers share the single operating system kernel. Moreover, it allows the processes of one container to access as many resources as possible. The virtual machine's startup and shutdown time is also very low. Therefore, it is very useful for the applications who requires frequent termination.

Virtualization Overhead. Memory management is one of the bottle neck for performance of the system. Therefore, cache and RAM are introduced to improve the performance of memory management unit. Virtualization technology creates the virtual environment of CPU, memory and network to run virtual machine. Virtual machine monitor uses binary translation technique to run the modified/unmodified operating system. It uses the technique of trap and emulate the instructions to execute in a non-privileged mode. CPU state and other primary structures of virtual machine are maintained by virtual machine monitor. Therefore, one page table of guest and one page table of host operating systems are working together to run the virtual machine's process [7]. As we know that virtual address consists of multiple levels like $L1$, $L2$, $L3$, $L4$ and *offset*. In order to get single physical address for virtual machine's process, it has to go through all the levels of host operating system or virtual machine monitor. This technique is called nested page table. Therefore, this technique considerably increases fetching time of the data. There are other overheads too like binary translation, trapping the privileged instructions and many more.

2.3 Cloud Computing

Cloud computing technology grows with considering computing as a utility [22]. Virtualization and high speed internet are the essential components of cloud computing. This technology gives new attractive platform to the hardware and software market. The research organizations or industries who has a batch of 100 jobs will not wait for 100 jobs serially to complete by single machine. They use parallel computing environment to reduce the execution time. There are three considerable cases with the computing infrastructure: underutilization, high utilization and peak demand for certain time. Instead of building and maintaining this expensive parallel environment for peak demand, resources can be leased from the cloud provider to reduce execution time as well as cost. Cloud computing is based on services like software, platform and infrastructure. The cloud is available on pay-as-you-go model. Therefore, cloud computing is a combination of software service and utility computing.

Infrastructure as a Service (IaaS) provides remote access of previously configured virtual machine. Providers also support scalability of these computing facilities on demand. IaaS cloud also offers storage service, firewall, IP addresses, virtual local area network and many software bundles for the user to access. Platform as a Service (PaaS) provides the access to the applications like database, web-server or programming language execution environment. User can develop and run the software without managing and maintaining the infrastructure and software. In the Software as a Service (SaaS) model, providers install and operate the software in the cloud and users can login and access these softwares. This model is used by the clients using pay-per-use basis.

There are three deployment models of cloud: private, public and hybrid. Private cloud is designed for single organization. Users from that specific organization can have access to this cloud. It can be managed by the same organization or a third party. There are couple of challenges to build the private cloud like security, capital investment and utilization of resources. Public cloud provides the service to all the customers from all over the world. *Amazon AWS* is an example of public cloud. Users can get the access through credit card and personal information. It is very flexible, less costly and scalable on demand. Hybrid cloud consists of two or more clouds. They can be a combination of private only or public only or private-public clouds. There are many challenges raised by hybrid cloud like, security between two clouds, interoperability of different clouds, virtual machine's costing mechanism and many more. In our testing environment, we have used OpenNebula and OpenStack as a cloud middleware to manage the virtual infrastructure.

2.4   OpenNebula

OpenNebula [23] is a cloud computing platform. It is used to manage the heterogeneous infrastructure in form of distributed data centers. It is a flexible solution with many features to manage enterprise data centers. The toolkit includes integration, security, accounting, management and scalability. OpenNebula offers to customers image catalogs, network catalogs, virtual machine template catalogs, virtual resources control and monitoring and multi-tier cloud application control and monitoring. High availability, storage, clusters, cloud bursting and application market is offered to the cloud operator.

OpenNebula supports complete virtual machine configuration with wide range of operating system, flexible network definition, automatic configuration of virtual machines and configuration of firewall. It also provides on-demand provision of virtual data centers (VDC) which is isolated virtual infrastructure where users and groups are under the control of the VDC administrator. Cloud bursting feature allows support for Amazon EC2. It is very easy to extend and integrate. We have used OpenNebula 4.4 with KVM as a hypervisor.

2.5   OpenStack

OpenStack [24] is also a platform to manage the cloud infrastructure. Open-Stack is started with the joint project of Rackspace and NASA in 2010. It is an open-source and free software framework under the term of Apache License. This framework is made up of series of interrelated projects specifically for storage, network, image and web-base dashboard. OpenStack compute service gives flexibility to offer on-demand computing resources which includes spawning, scheduling and decommissioning of virtual machines on demand. Horizon project is designed for web-based dashboard interface for the users for launching an instance, assigning IP addresses and configuring access controls. Image service is provided by glance project to store and retrieve the virtual machine disk image. For identity service, keystone project provides an authentication and authorization service for the users. We have used two node nova architecture for our implementation to test scientific jobs and standard benchmarks. We have used only basic services with KVM hypervisor for testing.

2.6   Scientific Workload

Scientific workload requires system size, high performance demand and execution model. Cloud computing technology gives a new opportunity to the scientist to test their tasks without building or maintaining highly configured infrastructure. Best characteristic of cloud computing for researcher is scalability on demand. Normally, scientific workload consists of read only (for analysis) or append only (to save the results) task. For read only jobs, they can access file from network or locally stored. In append only jobs, high I/O intensive work is to be done. Therefore, network, memory, I/O and CPU performance must be good to execute scientific jobs. Scientific workload consists of millions of tasks which can be divided into three types: high-performance computing, high-throughput computing and many-task computing. Figure 4 shows the jobs count of our site which varies from hundreds to thousands.

High-performance computing (HPC) is based on the performance of single job with its completion time. Therefore, it requires highly configured computing resources like supercomputer. Therefore, performance is measured in millions of instruction per second (MIPS) or in millions of floating-point operations per second (megaflops).

High-throughput computing (HTC) is described as the number of resources working on computational task for a long time. HTC is based on number of jobs complete per unit time. HTC provides robustness and reliability to the users. HTCondor and Portable Batch System (PBS) are the examples of HTC. HTC can support commodity hardware to execution of parallel task. High efficiency is

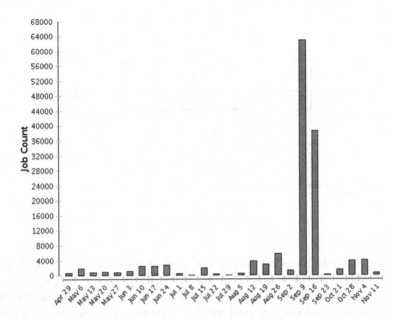

Fig. 4. Job count

not playing the main role in HTC computing environment. The main goal is to maximize the resources to be accessible for users to run many task together [25]. Performance is measured in number of jobs per hour.

Many-task computing (MTC) is a combination of high-throughput comput-ing and high-performance computing. In this case, resources are used to complete many computational tasks within a short time. MTC applications are loosely coupled or tightly coupled data or communication intensive. MTC can support loosely and tightly coupled jobs together on same system. MTC is allowing for finer grained tasks, independent task as well as ones with dependencies [26].

3   Experimental Setup

This section gives the detail description of the experimental setup. This setup is used to monitor the execution time of scientific workload as well as standard benchmarks. In this experiment, we have build three private clouds with Open-Nebula, OpenStack and XenServer. We also used public clouds like Amazon AWS, Google Engine and Rackspace.

3.1   Architecture

To create the private cloud, we have used commodity hardware of the configura-tion of Intel i7 processor with 4 GB RAM and 100 Mbps network. Figure 5 shows

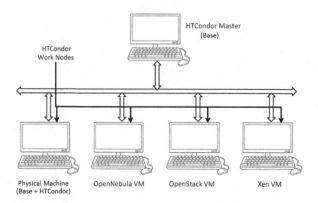

Fig. 5. Setup architecture

the architecture which we followed in this experiment. We have used HTCondor batch system software to submit the scientific jobs. HTCondor master is used to submit the scientific jobs which will be executed by work nodes. The work nodes consist of one physical machine (Base + HTCondor), OpenNebula virtual machines, OpenStack virtual machines and Xen virtual machines. Physical machine is used to compare the performance against virtual machines. HTCondor is installed on all the work nodes to get the scientific workload from HTCondor master. This setup helps to get the performance difference not only between physical machine and virtual machine, but also between private cloud and public cloud.

3.2 Virtual Machine Configuration

We have used equivalent environment of private as well as public cloud. We have selected Amazon, Google Engine and Rackspace as a public cloud. We also setup the private cloud with OpenNebula, OpenStack and XenServer. We have used KVM and Xen as a virtual machine monitor for the private cloud. Table 1 gives the configuration of virtual machines which are used in this experiment. Google Engine and Rackspace do not have virtual machine with configuration of 1 vCPU and 2 GB memory. Therefore, we have chosen the close configuration of other virtual machines. Some of the public clouds provide specialized virtual machines for CPU, memory and network intensive workload. However, we have chosen the general purpose machine because our private cloud is not configured specifically for any workload. About private cloud, we have written OTI (One Time Investment) under cost column of Table 1 which includes infrastructure and human resource cost to build and manage the private cloud.

4 Results

This section gives the performance evaluation of scientific workload as well as standard benchmarks. It also covers the discussion based on the results.

Table 1. Virtual machine configuration

Cloud name	vCPU	Memory	Hard Disk	Cloud type	Cost
OpenNebula	1	2	20	Private	OTI
OpenStack	1	2	20	Private	OTI
XenServer	1	2	20	Private	OTI
Amazon	1	2	20	Public	0.100 $/hr
Google Engine	1	1.7	20	Public	0.025 $/hr
Rackspace	1	1	20	Public	0.045 $/hr

4.1 Scientific Workload

We have characterized 4 different scientific jobs and executed on physical machine. Figure 6 shows the execution status of scientific jobs by using *top* command. Job with PID 3204 is compute intensive which is shown in the Fig. 6. PID 3277 is an I/O intensive job while PID 3080 is a memory intensive job. Network intensive job is PID 3254. These scientific jobs are developed in *python* language. Only network intensive job accesses the network to fetch the data. These results are taken on commodity hardware.

CPU Intensive. This job consists of matrix multiplication of two matrices. These matrices are changed each time and again compute the multiplication. This process repeats 9 million times to calculate final result. Therefore, this is CPU intensive job consuming 99 % CPU utilization as per Fig. 6.

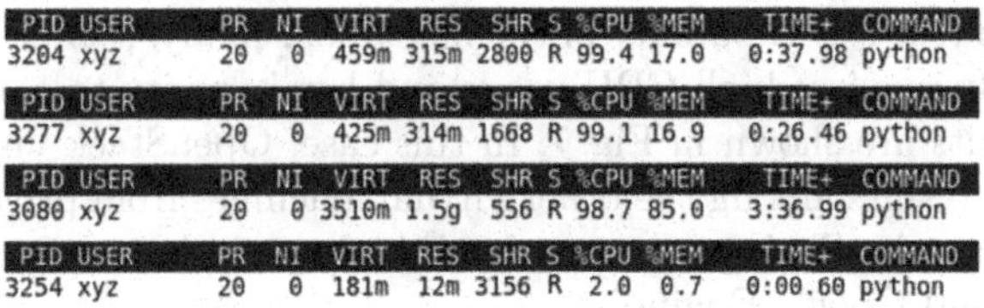

Fig. 6. Job execution status

Figure 7 shows the execution results where both the physical machines (Base and Base + HTCondor) gave the results within 45.5 min. While in the case of virtual machines, Xen virtual machine gave the result within 45.7 which is close to the physical machine. Same jobs are executed on public clouds, Amazon virtual machine took 132.6 min to calculate the result. Rackspace virtual machine gave the fastest result among the public clouds within 63.9 min.

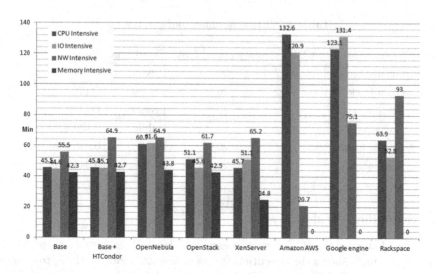

Fig. 7. Scientific workload execution time

Memory Intensive. This job creates new string and append another new string continuously. This job handles the huge string of the size 3.4 billion character. Therefore, it requires huge amount of memory which is shown in the Fig. 6. Figure 7 shows the results where Xen virtual machine surprisingly gave the results within 24.8 min. Xen virtual machine's execution time is even less than base (physical) machine. Virtual machines from public clouds are unable to give the results of these scientific jobs.

I/O Intensive. This job opens the file and appends it with different data. This job is created to test I/O capability of the machine. Finally, it ends with creation of 1.4 GB file. It requires high CPU usage and less memory to complete as per Fig. 6. The results are shown in Fig. 7. In this case, OpenStack virtual machine gave the fastest results among the other virtual machines from private clouds. In case of public clouds, Rackspace virtual machine gave the results quickly than Amazon AWS and Google Engine.

Network Intensive. This job downloads the file from specified location and creates local file by appending downloaded data. This job is designed to test the network capability of the virtual machine. Therefore, it requires very less CPU as well memory which is shown in the Fig. 6. Figure 7 shows the execution results. Amazon AWS virtual machine gave the fastest result of the network intensive job. It might be because of network bandwidth available in Amazon data center.

4.2 Standard Benchmark Execution

To evaluate the private and public cloud, we have chosen some of the standard benchmarks to understand more about performance difference.

Linux Kernel Compilation. In order to test the performance and throughput of the system, linux kernel compilation test is the best option. This test provides the performance by fast compiling single file and throughput by compiling multiple files within certain amount of time. We have chosen Linux Kernel 3.18 for this test. Figure 8 shows the compilation time for private and public cloud. Rackspace virtual machine compiles the kernel within 64.12 min while Google Engine virtual machine took 133.49 min to compile.

LMbench Benchmark. In cloud computing, main bottleneck for the performance is memory. In order to test the memory performance of private and public cloud virtual machines, we have used LMbench [28]. This benchmark consists of

- **Virtual memory system:** Creation of *mmaps* and *munmaps*, random read of *mmaped* file, and large transfers
- **File system:** Create and remove file, random I/O operation on small and large file, and measure *fsck* time.
- **Disks:** Small and large transfer of files
- **Networking:** Small and large transfer of packets
- **Processes:** Null process execution time, context switching

We have used 1 copy with capability of scheduler to place the jobs. 100 MB is used while testing with remaining default setting. Figure 8 shows the results of completion time of LMbench test. Amazon virtual machine is unable to execute this benchmark. While in the case of private cloud, OpenNebula virtual machine gave the fastest results among the private clouds.

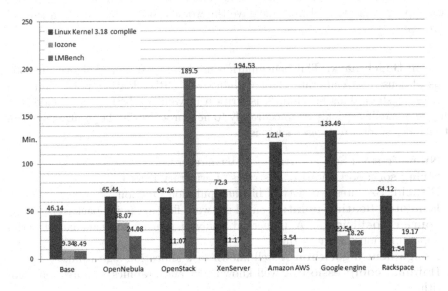

Fig. 8. Standard benchmark execution time

IOzone Benchmark. IOzone is a benchmark to test the filesystem [29]. This benchmark executes different types of file operation as follows.

- **Read:** read the existing file
- **Write:** creation of new file while maintaining its metadata. Maintaining a metadata leads to lower the performance.
- **Re-Read:** reading recently read file. This test is used the check cache performance.
- **Re-Write:** writing the already existing file. Performance is better than write because of presence of the metadata.
- **Random-Read/Write:** reading/ writing random location within a file. Performance depends on cache, number of disk and seek latency.
- **Backwards Read:** reading the file from bottom to top.
- **Strided Read:** This test reads 4 Kbytes and seek 200 Kbytes. It also repeats this pattern.
- **Fread/Fwrite:** reading and writing the file using *fread* and *fwrite* function respectively.
- **Mmap:** mapping the file in user's address space. Test is used to check the performance of *mmap()* for performing I/O.
- **Async I/O:** This test calculates the performance of POSIX async I/O for example *aio_read, aio_write* and *aio_error*.

In order to test all these test cases, we have used *./iozone -a* command. We have executed this benchmark on private as well as public cloud. This benchmark has capability to test variety of I/O operations. Figure 8 shows the time required to execute this benchmark. Rackspace virtual machine gave the best results among all other private as well as public cloud virtual machines. In this case, OpenNebula virtual machine gave the worst result with execution time of 38.07 min.

nbench Benchmark. This benchmark is used to test standard algorithms [30]. It is also used to elaborate the capabilities of CPU, floating point unit and memory system. This benchmark is written in vanilla ANSI C to provide best chance of moving quickly and accurately to new processor and operating system. It includes several test cases such as:

- **Numeric sort:** Sort an array of 32-bit integer.
- **String sort:** Sort an array of strings of arbitrary length.
- **Bitfield:** Executes a variety of bit manipulation function.
- **Emulated floating-point:** A small software floating-point package.
- **Fourier coefficients:** A numerical analysis routine for calculating series approximations of waveforms.
- **Assignment algorithm:** A well-known task allocation algorithm.
- **Haffman compression:** A well-known text and graphics compression algorithm.
- **IDEA encryption:** A relatively new block cipher algorithm.

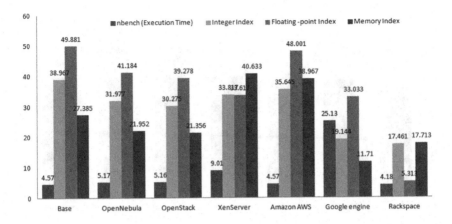

Fig. 9. nbench execution deatils

– **Neural Net:** A small but functional back-propagation network simulator.
– **LU Decomposition:** A robust algorithm for solving linear equation.

Figure 9 shows the memory, integer and floating point index of private and public cloud. It also shows the completion time of the test. Rackspace virtual machine gave the fastest results of this test. Even though, physical machine took 4.57 min to give the results. Google Engine virtual machine gave the worst results among all by spending 25.13 min.

Apache Benchmark. It is a single-threaded command line program to measure the performance of HTTP web server [27]. It is an open sources software which comes with Apache tools under the terms of Apache License. This benchmark shows the capability of serving these many requests per second. In order to improve the performance, Apache Bench can be run under multi-threaded environment. For this test, we have used single-threaded with the command *ab -n 1000 -c 5* http://www.facebook.com/. Figure 10 shows the performance of this benchmark on private and public cloud. Rackspace virtual machine gave the best results among other public cloud's virtual machines.

4.3   Discussion

As we know that the virtual machine comes with overhead. Table 2 shows the fastest and slowest virtual machine's execution time by normalizing with physical machine. For scientific jobs, public cloud shows the worst performance with increasing time of execution by two times or even three times. For CPU intensive job, physical machine gave the results within 45.5 min, while in the case of Amazon public cloud virtual machine gave the result in 132.6 min. Execution time is around three times longer than that of physical machine. In the case of nbench benchmark, Google Engine virtual machine's execution time is five times

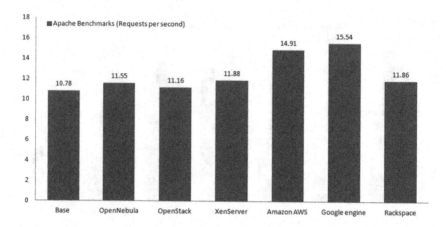

Fig. 10. Request per second

longer than that of physical machine. As per our observation, Rackspace virtual machine gave the best performance among other public clouds.

Table 2 shows the execution time with respect to physical machine. In the case of I/O intensive job, OpenStack virtual machine gave the result in 1.022 times of physical machine's execution time and Google Engine virtual machine's execution time is 2.95 times of physical machine's execution time. Therefore, virtual machine's execution time might be 1 to 3 times of physical machine's execution time.

For scientific computing, Global Science Experimental Data Hub Center (GSDC) [31] was supporting Collider Detector at Fermilab (CDF) [32] experiment by proving computational resources. CDF experiment is supported by

Table 2. Execution summary

Test	Fastest	Slowest
Scientific jobs		
CPU intensive	Xen (1.004)	Amazon (2.914)
Memory intensive	Xen (0.586)	all public cloud
I/O intensive	OpenStack (1.022)	GoogleEngine (2.95)
Network intensive	Amazon (0.373)	Rackspace (1.676)
Benchmarks		
Kernel compilation	Rackspace (1.389)	GoogleEngine (2.89)
LMbench	GoogleEngine (2.15)	Amazon
IOzone	Rackspace (0.164)	Amazon (2.413)
nbench	Rackspace (0.915)	GoogleEngine (5.5)
Apache Benchmark	OpenStack (1.035)	GoogleEngine (1.44)

number of Tier-1, Tier-2 and Tier-3 data centers. GSDC was the Tier-3 data center for CDF. GSDC allocated 1024 cores for CDF experiment which was running 24 × 7. Total number of jobs within 29 April to 11 November are 142078. The average execution time is 73.6 min. Therefore, total number of hours are around 174282. Figure 11 shows the execution status for 7 months.

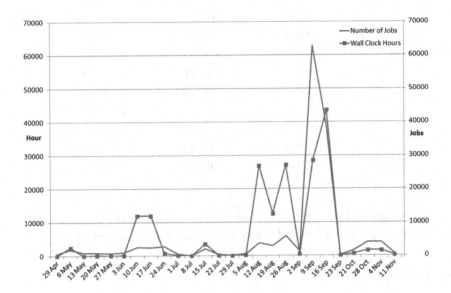

Fig. 11. CDF job history

If we consider execution time of CDF jobs as 1 (without virtual machine overhead), then estimated price required to execute CDF jobs on public cloud which is shown in first row of Table 3. Then, in the second row, execution time becomes 1.2 which includes 0.2 virtualization overhead. If we add the virtualization overhead, then the execution time increases and cost also increases. In our experiment, even though we have used commodity hardware to test the scientific jobs and standard benchmarks, still it took 5 times more than physical machine. If overhead goes up to 10 times then it requires around $174,282 to execute on Amazon for only 7 months as per Table 3. CDF project is running more than 30 years by supporting many data centers. Therefore, virtualization overhead is considerable point while making the decision of shifting grid computing to cloud computing. Table 3 shows the execution time with virtualization overhead and cost required for public cloud.

Power consumption of data center is also one of considerable problems in recent years. In order to solve this problem, researchers give a new solution, that is cloud computing. However, if the cloud computing increases the execution time then it needs to verify the effectiveness of cloud computing with respect to power consumption. Due to virtualization overhead, power consumption per job increases for private as well as public clouds even though power consumption

Table 3. Execution summary

Execution time	Amazon AWS $	Google Engine $	Rackspace $
1	17428.2	4357.05	7842.69
1.2	20913.84	5228.46	9411.228
1.5	26142.3	6535.575	11764.035
1.8	31370.76	7842.69	14116.842
2	34856.4	8714.1	15685.38
2.2	38342.04	9585.51	17253.918
2.5	43570.5	10892.625	19606.725
2.8	48798.96	12199.74	21959.532
3	52284.6	13071.15	23528.07
5	87141	21785.25	39213.45
8	139425.6	34856.4	62741.52
10	174282	43570.5	78426.9

per data center reduces. Therefore, we have to rediscover the relation between power consumption per job and power consumption per data center.

5   Related Work

This section describes the existing related work with respect to execution of scientific workload on cloud and the performance measurement of public cloud services based on different applications. Many-task scientific computing is tested with respect to performance [33]. In this paper, they have tested scientific workload on four different clouds, *Amazon EC2, GoGrid, ElasticHost* and *Mosso*. They have also used standard benchmarks and monitored the performance of each flavor of cloud computing. Their results showed that the cloud needs to improve in performance for scientific workload. This paper dose not consider private cloud to monitor the performance.

High performance computing application's feasibility is measured by Keith Jackson et al. [34]. This research work quantitatively examined the performance of HPC workload on *Amazon EC2* cloud. They found the relationship between performance and communication. As the communication increases the performance decreases. Therefore, they found that EC2 performance varies significantly because of shared nature of virtualization which is not suitable for high performance computing applications. This paper did not specify performance for high throughput computing jobs.

NASA Ames research center have hosted cloud platform, named as *Nebula*. They have tested the performance of high performance computing application on their private cloud platform. They found that, overall performance is reduced by 15 % to 48 % [35]. They also specified that virtualization overhead is about 10 %

to 25 %. Zheng Li et al. constructed the taxonomy for cloud related performance evaluation [36]. This taxonomy can be adopted for analyzing cloud computing performance and to design new infrastructure based on that evaluation. Another example of testing private cloud infrastructure is done by Ahmed Khurshid et al. at University of Illinois at Urbana-Champaign. They have used *PlanetLab* and *Emulab* to experiment on the private testbed [37]. They discovered that the cloud configuration and network characteristics affect the data transfer throughput.

Network I/O application's performance is studied by Yiduo Mei et al. [38]. If the virtual machines, which run network applications, co-located to on the same physical machine, then both of them are competing for same resource. During this scenario, the performance is degraded. Authors suggested that if providers avoid this then they can achieve over 40 % performance gain cloud consumers. Bogdan Nicolae tried to find the solution for cost effective cloud data storage [39]. This paper's goal is to reduce the storage space and bandwidth need to access the data with consideration of performance. The proposed method reduces the 40 % of storage space and required bandwidth.

Performance evaluation of cloud or virtualized resources are carried out by couple of other papers too [40–45] and many more. Most of them identified that the cloud computing technology faces the problem of virtualization overhead. None of the paper compares the results of private and public cloud together under same testing environment. Our contribution is to test the private and public cloud with scientific workload and standard benchmarks. We also showed that how this performance will affect on cost.

6 Conclusion

As the cloud computing is getting popular in the infrastructure market, many organizations and industries are attracted towards it. However, it needs to verify suitable type of workload for the cloud. In this paper, we tried to find out the reliability of cloud computing for scientific workload. Therefore, we started to monitor the resource consumption of scientific workload. Then, we categorized these jobs into CPU, memory, I/O and network intensive jobs. We measured the performance of scientific workload on private as well as public clouds including *Amazone AWS, Google Engine* and *Rackspace*. In order to elaborate more about performance, we have used the standard benchmarks like *LMbench, nbench, IOzone* and *Apache Benchmark* to execute on private and public cloud.

We found that the cloud computing technology needs considerable improvement for scientific workload. Virtualization layer enables overhead which leads to increase in execution time. Increased execution time leads to high power consumption of the data center. If we consider execution time of scientific jobs is around 1 month on a grid and cloud computing technology overhead is 0.5 times more than grid, then it will take around one and half month to execute the same set of jobs. Therefore, it depends on scientist that how quickly they want their results. In this paper, we also discuss the effectiveness of cloud computing for scientific workload, because the execution time and cost are directly proportion to each other. In case of power consumption, as the execution time increases the power consumption also increases irrespective of private or public cloud. Therefore, adoption of cloud computing may increase the power consumption per job. However, cloud computing technology gives new option for scientist to lease the opportunistic resources on demand.

Acknowledgment. This work was supported by the program of the Construction and Operation for Large-scale Science Data Center (K-16-L01-C06) and by National Research Foundation (NRF) of Korea (N-16-NM-CR01).

References

1. Mell, P., Grance, T.: The NIST definition of cloud computing. Commun. ACM **53**(6), 50 (2011)
2. Zhang, Q., Cheng, L., Boutaba, R.: Cloud computing: state-of-the-art and research challenges. J. Internet Serv. Appl. **1**(1), 7–18 (2010)
3. Foster, I., Zhao, Y., Raicu, I., Lu, S.: Cloud computing and grid computing 360-degree compared. In: Grid Computing Environments Workshop, GCE2008, pp. 1–10. IEEE (2008)
4. Mergen, M.F., Uhlig, V., Krieger, O., Xenidis, J.: Virtualization for high-performance computing. ACM SIGOPS Oper. Syst. Rev. **40**(2), 8–11 (2006)
5. Huber, N., von Quast, M., Hauck, M., Kounev, S.: Evaluating and modeling virtualization performance overhead for cloud environments. In: CLOSER, pp. 563–573 (2011)
6. McDougall, R., Anderson, J.: Virtualization performance: perspectives and challenges ahead. ACM SIGOPS Oper. Syst. Rev. **44**(4), 40–56 (2010)
7. Adams, K., Agesen, O.: A comparison of software and hardware techniques for x86 virtualization. ACM SIGPLAN Not. **41**(11), 2–13 (2006)
8. Barker, A., Hemert, J.: Scientific workflow: a survey and research directions. In: Wyrzykowski, R., Dongarra, J., Karczewski, K., Wasniewski, J. (eds.) PPAM 2007. LNCS, vol. 4967, pp. 746–753. Springer, Heidelberg (2008). doi:10.1007/978-3-540-68111-3_78
9. Yu, J., Buyya, R.: A taxonomy of scientific workflow systems for grid computing. ACM Sigmod Rec. **34**(3), 44–49 (2005)
10. Juve, G., Deelman, E., Vahi, K., Mehta, G., Berriman, B., Berman, B.P., Maechling, P.: Data sharing options for scientific workflows on Amazon EC2. In: Proceedings of the 2010 ACM/IEEE International Conference for High Performance Computing, Networking, Storage and Analysis, pp. 1–9. IEEE Computer Society (2010)
11. Koomey, J.G.: Estimating total power consumption by servers in the US and the world (2007)
12. Quang-Hung, N., Thoai, N., Son, N.T.: EPOBF: energy efficient allocation of virtual machines in high performance computing cloud. In: Hameurlain, A., Küng, J., Wagner, R., Dang, T.K., Thoai, N. (eds.) Transactions on Large-Scale Data- and Knowledge-Centered Systems XVI. LNCS, vol. 8960, pp. 71–86. Springer, Heidelberg (2014). doi:10.1007/978-3-662-45947-8_6
13. Greenberg, A., Hamilton, J., Maltz, D.A., Patel, P.: The cost of a cloud: research problems in data center networks. ACM SIGCOMM Comput. Commun. Rev. **39**(1), 68–73 (2008)

14. HTCondor, June 2015. http://research.cs.wisc.edu/htcondor/
15. Smith, I.C.: Experiences with running MATLAB applications on a power-saving condor pool. http://condor.liv.ac.uk/presentations/cardiff_condor.pdf
16. KVM, June 2015. http://www.linux-kvm.org/page/Main_Page
17. Kivity, A., Kamay, Y., Laor, D., Lublin, U., Liguori, A.: KVM: the Linux virtual machine monitor. Proc. Linux Symp. **1**, 225–230 (2007)
18. Chen, W., Lu, H., Shen, L., Wang, Z., Xiao, N., Chen, D.: A novel hardware assisted full virtualization technique. In: The 9th International Conference for Young Computer Scientists, ICYCS 2008, pp. 1292–1297. IEEE (2008)
19. Xen, June 2015. http://www.xenproject.org/
20. Barham, P., Dragovic, B., Fraser, K., Hand, S., Harris, T., Ho, A., Neugebauer, R., Pratt, I., Warfield, A.: Xen and the art of virtualization. ACM SIGOPS Oper. Syst. Rev. **37**(5), 164–177 (2003)
21. OpenVZ, June 2015. http://openvz.org/Main_Page
22. Armbrust, M., Fox, A., Griffith, R., Joseph, A.D., Katz, R., Konwinski, A., Lee, G., et al.: A view of cloud computing. Commun. ACM **53**(4), 50–58 (2010)
23. OpenNebula, June 2015. http://opennebula.org/
24. OpenStack, June 2015. https://www.openstack.org/
25. Livny, M., Basney, J., Raman, R., Tannenbaum, T.: Mechanisms for high throughput computing. SPEEDUP J. **11**(1), 36–40 (1997)
26. Raicu, I., Foster, I.T., Zhao, Y.: Many-task computing for grids and supercomputers. In: Workshop on Many-Task Computing on Grids and Supercomputers, MTAGS 2008, pp. 1–11. IEEE (2008)
27. Apache Benchmark, June 2015. http://httpd.apache.org/docs/2.2/programs/ab.html
28. LMbench, June 2015. http://www.bitmover.com/lmbench/
29. IOzone, June 2015. http://www.iozone.org/docs/
30. nbench, June 2015. http://www.tux.org/~mayer/linux/bmark.html
31. GSDC, June 2015. http://en.kisti.re.kr/supercomputing/
32. CDF, June 2015. http://www-cdf.fnal.gov/
33. Iosup, A., Ostermann, S., Yigitbasi, M.N., Prodan, R., Fahringer, T., Epema, D.H.J.: Performance analysis of cloud computing services for many-tasks scientific computing. IEEE Trans. Parallel Distrib. Syst. **22**(6), 931–945 (2011)
34. Jackson, K.R., Ramakrishnan, L., Muriki, K., Canon, S., Cholia, S., Shalf, J., Wasserman, H.J., Wright, N.J.: Performance analysis of high performance computing applications on the Amazon web services cloud. In: 2010 IEEE Second International Conference on Cloud Computing Technology and Science (CloudCom), pp. 159–168. IEEE (2010)
35. Saini, S., Heistand, S., Jin, H., Chang, J., Hood, R., Mehrotra, P., Biswas, R.: An application-based performance evaluation of NASA's nebula cloud computing platform. In: 2012 IEEE 9th International Conference on Embedded Software and Systems (HPCC-ICESS), 2012 IEEE 14th International Conference on High Performance Computing and Communication, pp. 336–343. IEEE (2012)
36. Li, Z., O'Brien, L., Cai, R., Zhang, H.: Towards a taxonomy of performance evaluation of commercial cloud services. In: 2012 IEEE 5th International Conference on Cloud Computing (CLOUD), pp. 344–351. IEEE (2012)
37. Khurshid, A., Al-Nayeem, A., Gupta, I.: Performance evaluation of the Illinois cloud computing testbed (2009)
38. Mei, Y., Ling L., Pu, X., Sivathanu, S.: Performance measurements and analysis of network i/o applications in virtualized cloud. In: 2010 IEEE 3rd International Conference on Cloud Computing (CLOUD), pp. 59–66. IEEE (2010)

39. Nicolae, B.: On the benefits of transparent compression for cost-effective cloud data storage. In: Hameurlain, A., Küng, J., Wagner, R. (eds.) Transactions on Large-Scale Data and Knowledge-Centered Systems III. LNCS, vol. 6790, pp. 167–184. Springer, Heidelberg (2011). doi:10.1007/978-3-642-23074-5_7
40. Diaz, C.O., Pecero, J.E., Bouvry, P., Sotelo, G., Villamizar, M., Castro, H.: Performance evaluation of an IaaS opportunistic cloud computing. In: 2014 14th IEEE/ACM International Symposium on Cluster, Cloud and Grid Computing (CCGrid), pp. 546–547. IEEE (2014)
41. Frey, S., Reich, C., Lthje, C.: Key performance indicators for cloud computing SLAs. In: The Fifth International Conference on Emerging Network Intelligence, EMERGING 2013, pp. 60–64 (2013)
42. Wang, H., Wang, F., Liu, J., Groen, J.: Measurement and utilization of customer-provided resources for cloud computing. In: 2012 Proceedings IEEE, INFOCOM, pp. 442–450. IEEE (2012)
43. Duy, T.V.T., Sato, Y., Inoguchi, Y.: Performance evaluation of a green scheduling algorithm for energy savings in cloud computing. In: 2010 IEEE International Symposium on Parallel and Distributed Processing, Workshops and Phd Forum (IPDPSW), pp. 1–8. IEEE (2010)
44. Stantchev, V.: Performance evaluation of cloud computing offerings. In: Third International Conference on Advanced Engineering Computing and Applications in Sciences, ADVCOMP 2009, pp. 187–192. IEEE (2009)
45. Jaikar, A., Dada, H., Kim, G.-R., Noh, S.-Y.: Priority-based virtual machine load balancing in a scientific federated cloud. In: 2014 IEEE 3rd International Conference on Cloud Networking (CloudNet), pp. 248–254. IEEE (2014)

Differential Erasure Codes for Efficient Archival of Versioned Data in Cloud Storage Systems

J. Harshan[✉], Anwitaman Datta, and Frédérique Oggier

Nanyang Technological University, Singapore, Singapore
jharshan@ntu.edu.sg

Abstract. In this paper, we study the problem of storing an archive of versioned data in a reliable and efficient manner. The proposed technique is relevant in cloud settings, where, because of the huge volume of data to be stored, distributed (scale-out) storage systems deploying erasure codes for fault tolerance is typical. However existing erasure coding techniques do not leverage redundancy of information across multiple versions of a file. We propose a new technique called differential erasure coding (DEC) where the differences (deltas) between subsequent versions are stored rather than the whole objects, akin to a typical delta encoding technique. However, unlike delta encoding techniques, DEC opportunistically exploits the sparsity (i.e., when the differences between two successive versions have few non-zero entries) in the updates to store the deltas using sparse sampling techniques applied with erasure coding. We first show that DEC provides significant savings in the storage size for versioned data whenever the update patterns are characterized by in-place alterations. Subsequently, we propose a practical DEC framework so as to reap storage size benefits against not just in-place alterations but also real-world update patterns such as insertions and deletions that alter the overall data sizes. We conduct experiments with several synthetic and practical workloads to demonstrate that the practical variant of DEC provides significant reductions in storage-overhead.

Keywords: Cloud storage · Backup and recovery · Fault tolerance · Erasure coded storage

1 Introduction

Over the last decade, cloud storage services have revolutionized the way we store and manage our digital data. Due to massive advances in the internet-, wireless-, and storage-technologies, plenty of file hosting services are nowadays offering storage and/or computing facilities to store huge amounts of data that are accessible from various locations on different devices and platforms. From an engineering view point, developing such ubiquitous cloud storage services necessitates in-depth understanding of, (i) *reliability:* how to store data across a network by guaranteeing a certain fault tolerance against failure of storage devices? (ii) *security:* how to protect data from security threats, both by a passive

© Springer-Verlag GmbH Germany 2016
A. Hameurlain et al. (Eds.): TLDKS XXX, LNCS 10130, pp. 23–65, 2016.
DOI: 10.1007/978-3-662-54054-1_2

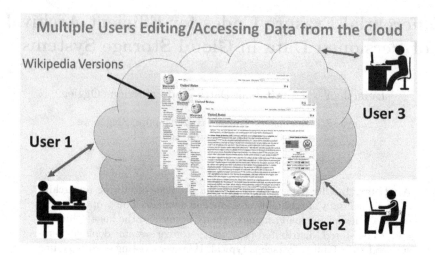

Fig. 1. Schematic depicting typical cloud storage application where multiple users access data from the cloud facility. In this illustrative figure, users are accessing and editing Wikipedia pages thereby creating a repository of multiple versions. Our work addresses a new framework of distributed storage systems for versioned data, aiming to lay foundations to new system architectures for cloud storage utilities supporting multiple versions.

adversary which is interested in reading the stored data, and also by an active one which is keen on manipulating the existing data? and (iii) *availability:* how to spread data across the network so as to speed up synchronization between client devices and the cloud. Each of the above aspects is a specialized area of study in this field, and all have been active areas of research.

In this work, we address the reliability aspect in cloud storage wherein we are interested in developing efficient ways of storing, accessing and modifying data in the cloud, while at the same time guaranteeing a certain level of fault tolerance against device (node) failures. In cloud storage networks, redundancy of the stored data is critical to ensure fault tolerance against node failures. While data replication remains a practical way of realizing this redundancy, the past years have witnessed the adoption of erasure codes for data archival, e.g. in Microsoft Azure [1], Hadoop FS [2], or Google File System [3], which offer a better trade-off between storage-overhead and fault tolerance. Thus, design of erasure coding techniques amenable to reliable and efficient storage has accordingly garnered a huge attention [5,9]. Once the reliability aspect of erasure codes for standalone objects is better understood, it is natural to question the reliability of versioned data. The need to store multiple versions of data arises in many scenarios. For instance, when editing and updating files, users may want to explicitly create a version repository using a framework like SVN [6]. Cloud based document editing or storage services also often provide the users access to older versions of the documents, e.g., Google Docs, Microsoft's Office 365, and Apple's iWork. See Fig. 1 for an illustrative example where multiple users access/modify data on Wikipedia thereby creating a versioned repository. Another scenario is that of system level back-up, where directories, whole file systems or databases are archived - and versions refer to the different system snapshots. Example systems include Dropbox which provides backup storage over which several users can collaborate creating multiple versions of data. In either of the two file centric settings, irrespective of whether a working copy used during editing is stored locally or on the cloud, or in a system level back up, say using copy-on-write [7], the back-end storage system needs to preserve the different versions reliably, and can leverage on erasure coding for reducing the storage-overheads.

1.1   Significance and Applications

It is well known that versioning systems get rid of duplicated contents across subsequent versions of a file. The main objective of versioning is to store only the changes from the preceding versions so as to reduce the overall storage size, and yet be able to accurately reconstruct any version requested by a user. In general, versioning concept falls within a broad topic of deduplication, that works on a plethora of files over space and time, and not only on temporal changes of a file. Meanwhile, in distributed storage systems erasure coding has gained enormous attention as it provides reduced storage-overhead when compared with the replication scheme. Although erasure coding schemes and architectures have been applied on standalone data objects in the past, the literature on erasure coding to versioned data is scarce. One possible reason for scarcity might be the possibility of a straightforward option to apply erasure coding on the changes (deltas), i.e., to treat versioning and erasure coding as independent entities. In this work, we explore a new direction to develop a close-coupled compression and erasure coding technique that can reduce the complexity of the versioning system and still yield high fault tolerance and significant storage gains. This work develops on our preliminary work in [18], where an erasure coding technique was proposed to reduce the I/O gains when retrieving multiple versions of data.

Applications for retrieving versioned archive include software development environments wherein multiple versions of modules are developed by different members of the project, and are often checked into the system at different time instants, e.g. management of software files over CVS. In such applications, the system administrator or the project/team lead would need to retrieve multiple versions at once in order to perform consistency checks and/or to possibly merge the contents based on the nature of changes. In back-up applications like "time machine"(where there is no revision history, etc. in contrast to a SVN like application), even if the user may eventually check out a few random versions to locate the version they want, it is often desirable to prefetch several subsequent versions so that the user can browse through them and also navigate consecutive versions to identify the one that is finally needed. There, our strategy will have a superior performance.

1.2 Related Works

Erasure codes have been extensively deployed in practical distributed storage systems for efficient and reliable storage of data [8]. Their choice over the standard replication technique comes as a natural course of action since erasure codes were proven to reduce the storage-overhead while maintaining a given fault tolerance level. However, in the recent past, with the objective of maintaining storage systems intact despite high failure rate of devices, [4], a plethora of new erasure code constructions have surfaced to not only reduce the storage-overhead but also facilitate low-complexity repair process for recovering lost data [1]. Since then most works have focused on distributed storage architectures for storing stand-alone objects, and not many have addressed the aspect of efficiently storing multiple versions of data. The topic of erasure coding for versioned data is loosely related to the issues of efficient updates [12–15], and of deduplication [16], which is the process of eliminating duplicate data blocks in order to eliminate unnecessary redundancy. Existing works on update of erasure coded data focus on the computational and communication efficiency in carrying out the updates, with the goal to store only the latest version of the data, and thus do not delve into efficient storage or manipulation of the previous versions. Recently, Wang and Cadambe [10] have addressed multi-version coding for distributed data, where the underlying problem is to encode different versions so that certain subsets of storage nodes can be accessed to retrieve the most common version among them. Their strategy has been shown applicable when the updates for the latest version do not reach all the nodes, possibly due to network problems. More recently, in [11], the authors have considered the problem of synchronizing data in storage networks under an edit model that includes deletions and insertions. They propose several variants of erasure codes that allow updates on the parity check values with low-bit rates and small storage-overhead. Apart from [10,11] that nearly touch upon the subject of storing versioned data, not many contributions exist in the literature that explicitly address erasure coding schemes for versioned data.

Capitalizing on the advances in erasure coding techniques for distributed storage, a straightforward option is to apply erasure coding on deltas of a versioning scheme. One such well-known scheme is *Rsync* [19], which is widely used for file transfer and synchronization across networks. The key idea behind Rsync is the rolling checksum computation, using which only the modified/new blocks between successive versions are transferred, thereby reducing the communication bandwidth. When such algorithms are applied to store versioned data, significant reduction in storage size is expected. Thus, Rsync scheme indirectly falls in the related works section of this topic. In [18] (extended abstract available in [17]), we have proposed erasure codes for storing multi-versioned data to benefit purely in terms of I/O, and for objects of fixed size. This paper develops on [18] to not only provide I/O benefits, but also to yield total storage savings when storing different versions of data. Furthermore, various system-level implementation issues of this work have been discussed in [24]. So this contribution distinguishes from [18,24] by exploring new erasure coding techniques that suit the underlying versioning model. In the next section, we summarize the key idea of this paper.

1.3 The Key Idea

Sparse Signal Recovery (SSR) [20] has garnered widespread applications as a powerful signal processing technique that can extract and store sparse signals with significantly fewer measurements than its conventional counterparts. In this paper, we explore how SSR ideas can be adapted by the storage community to facilitate reduced storage size for big data in cloud storage networks. In this context, the word *signal* refers to a vector of real numbers (with respect to some basis), whereas a *sparse signal* refers to one with fewer non-zero components compared to the length of the vector. Some applications of SSR include audio and video processing, medical imaging, and communication systems, where the signal acquisition process projects the desired sparse signal into a lower-dimensional space, and then appropriately recovers the higher-dimensional sparse signal from the lower-dimensional signal. Although SSR techniques are widely applied to signals over real numbers, this topic is also extendable to finite fields [21]. An illustrative example is discussed in Fig. 2. Now, the reader may ask, how does one obtain sparse vectors in storage systems? The answer lies in the fact that when different versions of data object are stored, there is a possibility of a user introducing few changes between subsequent versions, which in turn may result in sparse difference vectors.

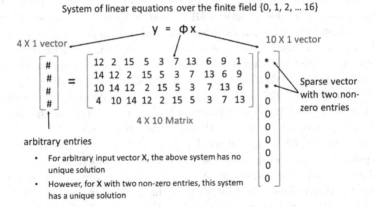

Fig. 2. An example to illustrate the possibility of recovering a sparse vector from a lower-dimensional vector. In this example, input vector which is 2-sparse (only two non-zero components irrespective of the positions) can be accurately recovered with 4 observations over the finite field $\mathbb{F}_{17} = \{0, 1, 2, \cdots, 16\}$, where arithmetic operations are over modulo 17.

1.4 Contributions

The contributions of this work are summarized below:

– We propose a new differential erasure coding (DEC) framework that falls under the umbrella of delta encoding techniques, where the differences (deltas) between subsequent versions are stored rather than the whole objects. The proposed technique exploits the sparsity in the differences among versions by applying techniques from sparse sampling [20], in order to reduce the storage-overhead (see Sects. 2 and 3). We have already proposed the idea of combining sparse sampling with erasure coding in [18], where we studied the benefits purely in terms of I/O, and for objects of fixed size. While retaining the combination of sparse sampling and erasure coding, this work introduces a different erasure coding strategy which provides storage-overhead benefits.
– We first present a simplistic layout of DEC that relies on fixed object lengths across successive versions of the data, so as to evaluate the right choice of erasure codes to store versioned data. We show that when all the versions are fetched in ensemble, there is also an equivalent gain in I/O operations. This comes at an increased I/O overhead when accessing individual versions. We accordingly propose some heuristics to optimize the basic DEC, and demonstrate that they ameliorate the drawbacks adequately without compromising the gains at all. Further, we show that the combination of sparse sampling and erasure coding yields other practical benefits such as the possibility of employing fewer erasure codes against different sparsity levels of the update patterns. (see Sect. 4).
– In the later part of this paper, we extend the preliminary ideas of DEC to develop a framework for practical DEC that is robust to real-world update patterns across versions such as insertions and deletions which may alter the overall size of the data object. Along that direction, we acknowledge that insertions and deletions may ripple changes across the object at the coding granularity, and may also increase the object size. Such rippling effect could in particular render DEC useless, and obliterate the consequent benefits. We apply the zero-padding idea introduced in [24] to ameliorate the aforementioned problems of insertions and deletions (see Sect. 5).
– For storing versioned data, the total storage size for *deltas* inclusive of zero pads (prior to erasure coding) is used as the metric to evaluate the quantum and placement of zero pads against a wide range of workloads that include insertions and deletions, both bursty and distributed in nature (see Sect. 6). We compare the storage savings offered by the practical DEC technique with the standard baselines against both synthetic and practical datasets. The baselines include (i) a non-differential scheme, where different versions of a data object are encoded in full without exploiting the sparsity across them, (ii) selective encoding scheme, wherein only modified blocks between the versions are stored, (iii) Rsync, a delta encoding technique for file transfer and synchronization across networks, and (iv) gz compression algorithm applied on individual versions to reduce the storage size of each version. Among the

four baselines, we show that DEC outperforms (i), (ii) and (iv), in terms of storage size, while trades-off storage size to computational complexity when compared with (iii). We use Wikipedia datasets to showcase the impact of DEC scheme on practical datasets.

1.5 System Model for Version Management

Any digital content to be stored, be it a file, directory, database, or a whole file system, is divided into data chunks, shown as phase ① in Fig. 3. The proposed coding techniques are agnostic of the nuances of the upper levels, and all subsequent discussions will be at the granularity of these chunks, which we will refer to as data objects or just objects.

Formally, we denote by $\mathbf{x} \in \mathbb{F}_q^k$ a data object to be stored over a network, that is, the object is seen as a vector of k blocks (phase ②) taking value in the alphabet $\mathbb{F}_q$, with $\mathbb{F}_q$ the finite field with q elements, q a power of 2 typically. Encoding for archival of an object $\mathbf{x}$ across n nodes is done (phase ③) using an (n, k) linear code, that is $\mathbf{x}$ is mapped to the codeword

$$\mathbf{c} = \mathbf{Gx} \in \mathbb{F}_q^n, \ n > k, \tag{1}$$

for $\mathbf{G}$ an $n \times k$ generator matrix with coefficients in $\mathbb{F}_q$. We use the term *systematic* to refer to a codeword $\mathbf{c}$ whose k first components are $\mathbf{x}$, that is $c_i = x_i$, $i = 1, \ldots, k$. This described what is a standard encoding procedure used in erasure coding based storage systems. We suppose next that the content mutates, and we wish to store all the versions.

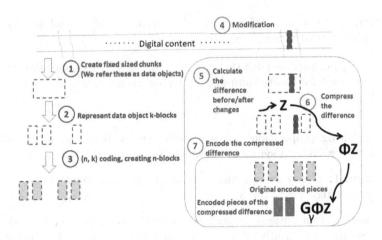

Fig. 3. An overview of the coding strategy using compressed differences

Let $\mathbf{x}_1 \in \mathbb{F}_q^k$ denote the first version of a data object to be stored. When it is modified (phase ④), a new version $\mathbf{x}_2 \in \mathbb{F}_q^k$ of this object is created. More generally, a new version $\mathbf{x}_{j+1}$ is obtained from $\mathbf{x}_j$ to produce over time a sequence $\{\mathbf{x}_j \in \mathbb{F}_q^k, \ j = 1, 2, \ldots, L < \infty\}$ of L different versions of a data object, to be stored in the network. We are not concerned with the application level semantic of the modifications, but with the bit level changes in the object. Thus the changes between two successive versions are captured by the relation

$$\mathbf{x}_{j+1} = \mathbf{x}_j + \mathbf{z}_{j+1}, \tag{2}$$

where $\mathbf{z}_{j+1} \in \mathbb{F}_q^k$ denotes the modifications (in phase ⑤) of the jth update. We first assume fixed object lengths across successive versions of data so as to build an uncomplicated framework for the differential strategy. Such a framework shields us from unnecessarily delving into system specificities, instead, serves as a foundation to evaluate various erasure coding techniques to store multiple versions of data. We show that the design, analysis and assessment of the coding techniques are oblivious to the nuances of how the data object is broken down into several chunks prior to the encoding purposes, thereby facilitating us to segregate *chunk synthesis* and *erasure coding* blocks as two independent entities. In the later part of this work (see Sect. 5), we discuss how to relax the fixed object length assumption and yet develop a practical DEC scheme, that is robust to variable object lengths across successive versions.

The key idea is that when the changes from $\mathbf{x}_j$ to $\mathbf{x}_{j+1}$ are small (decided by the sparsity of $\mathbf{z}_{j+1}$), it is possible to apply sparse sampling [20], which permits to represent a k-length γ-sparse vector $\mathbf{z}$ (see Definition 1) with less than k components (phase ⑥) through a linear transformation on $\mathbf{z}$, which does not depend on the position of the non-zero entries, in order to gain in storage efficiency.

Definition 1. *For some integer* $1 \leq \gamma < k$, *a vector* $\mathbf{z} \in \mathbb{F}_q^k$ *is said to be* γ-*sparse if it contains at most* γ *non-zero entries.*

Let $\mathbf{z} \in \mathbb{F}_q^k$ be γ-sparse such that $\gamma < \frac{k}{2}$, and $\Phi \in \mathbb{F}_q^{2\gamma \times k}$ denote the *measurement matrix* used for sparse sampling. The compressed representation $\mathbf{z}' \in \mathbb{F}_q^{2\gamma}$ of $\mathbf{z}$ is obtained as

$$\mathbf{z}' = \Phi \mathbf{z}. \tag{3}$$

The following proposition[1] gives a sufficient condition on Φ to uniquely recover $\mathbf{z}$ from $\mathbf{z}'$ using a syndrome decoder [21, Sect. II.B].

Proposition 1. *If any* 2γ *columns of* Φ *are linearly independent, the* γ-*sparse vector* $\mathbf{z}$ *can be recovered from* $\mathbf{z}'$.

Once sparse modifications are compressed, which reduces the I/O reads, they are encoded into codewords of length $< n$ (phase ⑦) decreasing in turn the storage-overhead.

---

[1] The proof for the proposition follows from the property that any 2γ columns of a parity check matrix of a linear code with minimum distance $2\gamma + 1$ are linearly independent.

2   Differential Erasure Encoding for Version-Control

Let $\{\mathbf{x}_j \in \mathbb{F}_q^k, \ 1 \leq j \leq L\}$ be the sequence of versions of a data object to be stored. The changes from $\mathbf{x}_j$ to $\mathbf{x}_{j+1}$ are reflected in the vector $\mathbf{z}_{j+1} = \mathbf{x}_{j+1} - \mathbf{x}_j$ in (2) which is γ_{j+1}-sparse (see Definition 1) for some $1 \leq \gamma_{j+1} \leq k$. The value γ_{j+1} may a priori vary across versions of one object, and across application domains. All the versions $\mathbf{x}_1, \ldots, \mathbf{x}_L$ need protection from node failures, and are archived using a linear erasure code (see (1)).

2.1   Object Encoding

We describe a generic *differential* encoding (called **Step** $j+1$) suited for efficient archival of versioned data, which exploits the sparsity of the updates, when $\gamma_{j+1} < \frac{k}{2}$, to reduce the storage-overheads of archiving all the versions reliably. We assume that one storage node is in possession of two versions, say $\mathbf{x}_j$ and $\mathbf{x}_{j+1}$ of one data object, $j = 1, \ldots, L - 1$. The corresponding implementation is discussed in Subsect. 2.2.

Step $j + 1$. For the two versions $\mathbf{x}_j$ and $\mathbf{x}_{j+1}$, the difference vector $\mathbf{z}_{j+1} = \mathbf{x}_{j+1} - \mathbf{x}_j$ and the corresponding sparsity level γ_{j+1} are computed. If $\gamma_{j+1} \geq \frac{k}{2}$, the object $\mathbf{z}_{j+1}$ is encoded as $\mathbf{c}_{j+1} = \mathbf{G}\mathbf{z}_{j+1}$. On the other hand, if $\gamma_{j+1} < \frac{k}{2}$, then $\mathbf{z}_{j+1}$ is first compressed (see (3)) as

$$\mathbf{z}'_{j+1} = \Phi_{\gamma_{j+1}} \mathbf{z}_{j+1},$$

where $\Phi_{\gamma_{j+1}} \in \mathbb{F}_q^{2\gamma_{j+1} \times k}$ is a measurement matrix such that any $2\gamma_{j+1}$ of its columns are linearly independent (see Proposition 1). Subsequently, $\mathbf{z}'_{j+1}$ is encoded as

$$\mathbf{c}_{j+1} = \mathbf{G}_{\gamma_{j+1}} \mathbf{z}'_{j+1},$$

where $\mathbf{G}_{\gamma_{j+1}} \in \mathbb{F}_q^{n_{\gamma_{j+1}} \times 2\gamma_{j+1}}$ is the generator matrix of an $(n_{\gamma_{j+1}}, 2\gamma_{j+1})$ erasure code with storage-overhead κ. The components of $\mathbf{c}_{j+1}$ are distributed across a set $\mathcal{N}_{j+1}$ of $n_{\gamma_{j+1}}$ nodes, whose choice is discussed in Subsect. 2.2.

Since γ_{j+1} is random, a total of $\lceil \frac{k}{2} \rceil$ erasure codes denoted by

$$\mathcal{G} = \{\mathbf{G}, \mathbf{G}_1, \ldots, \mathbf{G}_{\lceil \frac{k}{2} \rceil - 1}\},$$

and a total of $\lceil \frac{k}{2} \rceil - 1$ measurement matrices denoted by $\Sigma = \{\Phi_1, \Phi_2, \ldots, \Phi_{\lceil \frac{k}{2} \rceil - 1}\}$ have to be designed a priori. The erasure codes may be taken systematic and/or MDS (that is, such that any $n - k$ failure patterns are tolerated), our scheme works irrespective of these choices. This encoding strategy implies one extra matrix multiplication whenever a sparse difference vector is obtained.

We give a toy example to illustrate the computations.

```
1: procedure ENCODE(𝒳, 𝒢, Σ)
2:     FOR 0 ≤ j ≤ L − 1
3:         IF j = 0
4:             return c₁ = Gx₁;
5:         ELSE (This part summarizes Step j + 1 in the text)
6:             Compute zⱼ₊₁ = xⱼ₊₁ − xⱼ;
7:             Compute γⱼ₊₁;
8:             IF γⱼ₊₁ ≥ k/2
9:                 return cⱼ₊₁ = Gzⱼ₊₁;
10:            ELSE
11:                Compress zⱼ₊₁ as z'ⱼ₊₁ = Φγⱼ₊₁ zⱼ₊₁;
12:                return cⱼ₊₁ = Gγⱼ₊₁ zⱼ₊₁;
13:            END IF
14:        END IF
15:    END FOR
16: end procedure
```

Fig. 4. Encoding procedure for DEC

Example 1. Take $k = 4$, suppose that the digital content is written in binary as (100110010010) and that the linear code used for storage is a $(6, 4)$ code over $\mathbb{F}_8$. To create the first data object x_1, cut the digital content into $k = 4$ chunks 100, $110, 010, 010$, so that x_1 is written over $\mathbb{F}_8$ as $x_1 = (1, 1+w, w, w)$ where w is the generator of $\mathbb{F}_8^*$, satisfying $w^3 = w + 1$. The next version of the digital content is created, say (10011011001). Similarly x_2 becomes $x_2 = (1, 1 + w, 1 + w, w)$, and the difference vector z_2 is given by $z_2 = x_2 - x_1 = (0, 0, 1, 0)$, with $\gamma_2 = 1 < k/2$. Apply a measurement matrix $\Phi_{\gamma_2} = \Phi_1$ to compress z_2:

$$\Phi_1 z_2 = \begin{bmatrix} 1 & 0 & w & w+1 \\ 0 & 1 & w+1 & w \end{bmatrix} \begin{bmatrix} 0 \\ 0 \\ 1 \\ 0 \end{bmatrix} = \begin{bmatrix} w \\ w+1 \end{bmatrix} = z_2'.$$

Note that every two columns of Φ_1 are linearly independent (see Proposition 1), thus allowing the compressed vector to be recovered. Encode z_2' using a single parity check code:

$$c_2 = \begin{bmatrix} 1 & 0 \\ 0 & 1 \\ 1 & 1 \end{bmatrix} \begin{bmatrix} w \\ w+1 \end{bmatrix} = \begin{bmatrix} w \\ w+1 \\ 1 \end{bmatrix}.$$

2.2   Implementation and Placement

Caching. To store x_{j+1} for $j \geq 1$, the proposed scheme requires the calculation of differences between the existing version x_j and the new version x_{j+1} in (2). However, it does not store x_j, but x_1 together with $z_2, \ldots, z_j$. Reconstructing x_j before computing the difference and encoding the new difference is expensive in terms of I/O operations, network bandwidth, latency as well as computations. A practical remedy is thus to cache a full copy of the latest version $\mathbf{x}_j$, until a new version $\mathbf{x}_{j+1}$ arrives. This also helps in improving the response time and overheads of data read operations in general, and thus disentangles the system performance from the storage efficient resilient storage of all the versions.

Considering caching as a practical method, an algorithm summarizes the differential erasure coding (DEC) procedure in Fig. 4. The input and the output of the algorithm are $\mathcal{X} = \{\mathbf{x}_j \in \mathbb{F}_q^k, \ 1 \le j \le L\}$ and $\{\mathbf{c}_j, \ 1 \le j \le L\}$, respectively.

Placement Consideration. The choice of the sets $\mathcal{N}_{j+1}$, $j = 0, \ldots, L - 1$ of nodes over which the different versions are stored needs a closer introspection. Since $\mathbf{x}_1$ together with $\mathbf{z}_2, \ldots, \mathbf{z}_j$ are needed to recover $\mathbf{x}_j$ (see also Subsect. 2.4), if $\mathbf{x}_1$ is lost, $\mathbf{x}_j$ cannot be recovered, and thus there is no gain in fault tolerance by storing $\mathbf{x}_j$ in a different set of nodes than $\mathcal{N}_1$. Furthermore, since $n_{\gamma_j} < n$, codewords $\mathbf{c}_i$s may have different resilience to failures. The dependency of $\mathbf{x}_j$ on previous versions suggests that the fault-tolerance of subsequent versions are determined by the worst fault-tolerance achieved among $\mathbf{c}_i$s for $i < j$.

Example 2. We continue Example 1, where $\mathbf{x}_1$ is encoded into $\mathbf{c}_1 = (c_{11}, \ldots, c_{16})$ using a $(6, 4)$ MDS code. Allocate c_{1i} to N_i, that is use the set $\mathcal{N}_1 = \{N_1, \ldots, N_6\}$ of nodes. Store $\mathbf{c}_2$ in $\mathcal{N}_2 = \{N_1, N_2, N_3\} \subset \mathcal{N}_1$ for collocated placement, and in $\mathcal{N}_2 = \{N_1', N_2', N_3'\}$, $\mathcal{N}_2 \cap \mathcal{N}_1 = \emptyset$ for distributed placement. Let p be the probability that a node fails, and failures are assumed independent. We compute the probability to recover both $\mathbf{x}_1$ and $\mathbf{x}_2$ in case of node failures (known as *static resilience*) for both distributed and collocated strategies.

For distributed placement, the set of error events for losing $\mathbf{x}_1$ is $\mathcal{E}_1 = \{3 \text{ or more nodes fail in } \mathcal{N}_1\}$. Hence, the probability $\text{Prob}(\mathcal{E}_1)$ of losing $\mathbf{x}_1$ is given by

$$p^6 + C_5^6 p^5 (1 - p) + C_4^6 p^4 (1 - p)^2 + C_3^6 p^3 (1 - p)^3, \tag{4}$$

where C_r^m denotes the m choose r operation. The set of error events for losing $\mathbf{z}_2$ stored with a $(3,2)$ MDS code is $\mathcal{E}_2 = \{2 \text{ or } 3 \text{ nodes fail in } \mathcal{N}_2\}$. Thus, $\mathbf{z}_2$ is lost with probability

$$\text{Prob}(\mathcal{E}_2) = p^3 + C_2^3 p^2 (1 - p). \tag{5}$$

From (4) and (5), the probability of retaining both versions is

$$\text{Prob}_d(\mathbf{x}_1, \mathbf{x}_2) \triangleq (1 - \text{Prob}(\mathcal{E}_1))(1 - \text{Prob}(\mathcal{E}_2)). \tag{6}$$

The set of error events for losing $\mathbf{x}_1$ or $\mathbf{z}_2$ is

$$\mathcal{E}_1 \cup \mathcal{E}_2 = \{3 \text{ or more nodes fail}\} \cup \{\text{specific 2 nodes failure}\}$$

for collocated placement. Out of C_2^6 possible 2 node failure patterns, 3 patterns contribute to the loss of the object $\mathbf{z}_2$. Therefore, $\text{Prob}(\mathcal{E}_1 \cup \mathcal{E}_2)$ is

$$p^6 + C_5^6 p^5 (1 - p) + C_4^6 p^4 (1 - p)^2 + C_3^6 p^3 (1 - p)^3 + 3 p^2 (1 - p)^4$$

from which, the probability of retaining both the versions is

$$\text{Prob}_c(\mathbf{x}_1, \mathbf{x}_2) \triangleq 1 - \text{Prob}(\mathcal{E}_1 \cup \mathcal{E}_2). \tag{7}$$

In Fig. 5, we compare (6) and (7) for different values of p from 0.001 to 0.05. The plot shows that collocated allocation results in better resilience than the distributed case.

Optimized Step $j+1$. Based on these insights, a practical change of **Step** j is: if $\gamma_{j+1} \geq \frac{k}{2}$, $\mathbf{z}_{j+1}$ is discarded and $\mathbf{x}_{j+1}$ is encoded as $\mathbf{c}_{j+1} = \mathbf{G}\mathbf{x}_{j+1}$, to ensure that a whole version is again encoded. Since many contiguous sparse versions may be created, we put as a heuristic an iteration threshold ι, after which even if all differences from one version to another stay very sparse, a whole version is used for coding and storage.

2.3   On the Storage-Overhead

Since employed erasure codes depend on the sparsity level, the storage-overhead of the above differential encoding improves upon that of encoding different versions independently. The average gains in storage-overhead are discussed in Subsect. 3.3. Formally, the total storage size till the l-th version is

$$\delta(\mathbf{x}_1, \mathbf{x}_2, \ldots, \mathbf{x}_l) = n + \sum_{j=2}^{l} \min(2\kappa\gamma_j, n) \leq ln,$$

for $2 \leq l \leq L$. The storage-overhead for the **Optimized Step** $j+1$ is the same as that of **Step** $j+1$ since for $\gamma_{j+1} \geq \frac{k}{2}$, the coded objects $\mathbf{G}\mathbf{x}_{j+1}$ and $\mathbf{G}\mathbf{z}_{j+1}$ have the same size.

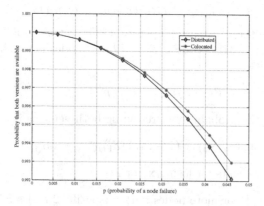

Fig. 5. Placement consideration: comparing probability that both versions are available

2.4 Object Retrieval

Suppose that L versions of a data object are archived using **Step** $j + 1$, $j \leq L - 1$ and the user needs to retrieve some $\mathbf{x}_l$, $1 < l \leq L$. Assuming that there are enough encoded blocks for each $\mathbf{c}_i$ ($i \leq l$) available, relevant nodes in the sets $\mathcal{N}_1, \ldots, \mathcal{N}_l$ are accessed to fetch and decode the $\mathbf{c}_i$ to obtain $\mathbf{x}_1$, and the $l - 1$ compressed differences $\mathbf{z}'_2, \mathbf{z}'_3, \ldots, \mathbf{z}'_l$. See Subsect. 2.2 for a discussion on placement and an illustration that reusing the same set of nodes gives the best availability with MDS codes, hence bounding the number of accessed nodes by $|\mathcal{N}_1|$. All compressed differences sharing the same sparsity can be added first, and then decompressed, since

$$\sum_{i \in J_\gamma} \mathbf{z}'_i = \Phi_\gamma \sum_{i \in J_\gamma} \mathbf{z}_i$$

for $J_\gamma = \{j | \gamma_j = \gamma\}$. The cost of recovering $\sum_{i \in J_\gamma} \mathbf{z}_i$ is only one decompression instead of $|J_\gamma|$, with which $\mathbf{x}_l$ is given by

$$\mathbf{x}_l = \mathbf{x}_1 + \sum_{j=2}^{l} \mathbf{z}_j.$$

A minimum of k I/O reads is needed to retrieve $\mathbf{x}_1$. For $\mathbf{z}_j$ ($2 \leq j \leq l$), the number of I/O reads may be lower than k, depending on the update sparsity. If $\gamma_j < \frac{k}{2}$, then $\mathbf{z}'_j$ is retrieved with $2\gamma_j$ I/O reads, while if $\gamma_j \geq \frac{k}{2}$, then $\mathbf{z}_j$ is recovered with k I/O reads, so that $\min(2\gamma_j, k)$ I/O reads are needed for $\mathbf{z}_j$. The total number of I/O reads to retrieve $\mathbf{x}_l$ is

$$\eta(\mathbf{x}_l) = k + \sum_{j=2}^{l} \min(2\gamma_j, k) \tag{8}$$

and so is the total number of I/O reads to retrieve the first l versions: $\eta(\mathbf{x}_1, \mathbf{x}_2, \ldots, \mathbf{x}_l) = \eta(\mathbf{x}_l)$.

To retrieve $\mathbf{x}_l$ for $1 \leq l \leq L$, when archival was done using **Optimized Step** $j + 1$, $j \leq L - 1$, look for the most recent version $\mathbf{x}_{l'}$ such that $l' \leq l$ and $\gamma_{l'} \geq \frac{k}{2}$. Then, using $\{\mathbf{x}_{l'}, \mathbf{z}_{l'+1}, \ldots, \mathbf{z}_l\}$, the object $\mathbf{x}_l$ is reconstructed as $\mathbf{x}_l = \mathbf{x}_{l'} + \sum_{j=l'+1}^{l} \mathbf{z}_j$. Hence, the total number of I/O reads is

$$\eta(\mathbf{x}_l) = k + \sum_{j=l'+1}^{l} \min(2\gamma_j, k). \tag{9}$$

The number of I/O reads to retrieve the first l versions is the same as for **Step** $j + 1$.

Example 3. Assume that $L = 20$ versions of an object of size $k = 10$ are differentially encoded, with sparsity profile

$$\{\gamma_j, \ 2 \leq j \leq L\} = \{3, 8, 3, 6, 7, 9, 10, 6, 2, 2, 3, 9, 3, 9, 3, 10, 4, 2, 3\}.$$

The storage pattern is $\{\mathbf{x}_1, \mathbf{z}_2, \mathbf{z}_3, \ldots, \mathbf{z}_{20}\}$. Assuming $\mathbf{x}_1$ is not sparse, the I/O read numbers to access $\{\mathbf{x}_1, \mathbf{z}_2, \mathbf{z}_3, \ldots, \mathbf{z}_{20}\}$ are

$$\{10, 6, 10, 6, 10, 10, 10, 10, 10, 4, 4, 6, 10, 6, 10, 6, 10, 8, 4, 6\}.$$

The total I/O reads to recover all the 20 versions is 156 (instead of 200 for the non-differential method). The total storage space for all the 20 versions assuming a storage-overhead of 2 is 312 (instead of 400 otherwise). The I/O read numbers to recover $\{\mathbf{x}_1, \mathbf{x}_2, \mathbf{x}_3, \ldots, \mathbf{x}_{20}\}$ are

$$\{10, 16, 26, 32, 42, 52, 62, 72, 82, 86, 90, 96, 106, 112, 122, 128, 138, 146, 150, 156\},$$

while for the optimized step, we get $\{10, 16, 10, 16, 10, 10, 10, 10, 10, 14, 18, 24, 10, 16, 10, 16, 10, 18, 22, 28\}$.

3   Reverse Differential Erasure Coding

In Table 1, we summarize the total storage size and the number of I/O reads required by the (forward) differential method. If some γ_j, $1 \leq j \leq l$, are smaller than $\frac{k}{2}$, then the number of I/O reads for joint retrieval of all the versions $\{\mathbf{x}_1, \mathbf{x}_2, \ldots, \mathbf{x}_l\}$ is lower than that of the traditional method. However, this advantage comes at the cost of higher number of I/O reads for accessing the l-th version $\mathbf{x}_l$ alone. Therefore, for applications where the latest archived versions are more frequently accessed than the joint versions, the overhead for reading the latest version dominates the advantage of reading multiple versions. For such applications, we apply a variant of the differential method called the reverse DEC, wherein the order of storing the difference vectors is reversed [6].

Table 1. I/O access metrics for the traditional and the differential schemes to store $\{\mathbf{x}_1, \mathbf{x}_2, \ldots, \mathbf{x}_l\}$

Parameter	Traditional	Forward differential	Reverse differential
I/O reads to read the l-th version	k	$k + \sum_{j=2}^{l} \min(2\gamma_j, k)$	k
I/O reads to read the first l-th versions	lk	$k + \sum_{j=2}^{l} \min(2\gamma_j, k)$	$k + \sum_{j=2}^{l} \min(2\gamma_j, k)$
Number of Encoding operations	1 (on the version)	1 (on the latest version)	2 (on the latest and the preceding version)
Total Storage Size till the l-th version	ln	$n + \sum_{j=2}^{l} \min(2\kappa\gamma_j, n)$	$n + \sum_{j=2}^{l} \min(2\kappa\gamma_j, n)$

3.1   Object Encoding

As in Subsect. 2.1, we assume that one node stores the latest version $\mathbf{x}_j$ and the new version $\mathbf{x}_{j+1}$ of a data object. Since $\mathbf{x}_j$ is readily obtained, caching is less critical here.

Step $j+1$. Compute the difference vector $\mathbf{z}_{j+1} = \mathbf{x}_{j+1} - \mathbf{x}_j$ and its sparsity level γ_{j+1}. The object $\mathbf{x}_{j+1}$ is encoded as $\mathbf{c}_{j+1} = \mathbf{G}\mathbf{x}_{j+1}$ and stored in $\mathcal{N}_{j+1}$. Furthermore, if $\gamma_{j+1} < \frac{k}{2}$, then $\mathbf{z}_{j+1}$ is first compressed as $\mathbf{z}'_{j+1} = \Phi_{\gamma_{j+1}}\mathbf{z}_{j+1}$, and then encoded as $\mathbf{c} = \mathbf{G}_{\gamma_{j+1}}\mathbf{z}'_{j+1}$, where $\mathbf{G}_{\gamma_{j+1}}$ is the generator matrix of an $(n_{\gamma_{j+1}}, 2\gamma_{j+1})$ erasure code. Finally, the preceding version $\mathbf{c}_j$ is overwritten as $\mathbf{c}_j = \mathbf{c}$.

A key feature is that in addition to encoding the latest version $\mathbf{x}_{j+1}$, the preceding version is also re-encoded depending on the sparsity level γ_{j+1}, resulting in two encoding operations (instead of one for the method in Subsect. 2.1).

A summary of the encoding is provided in Fig. 6. The storage-overhead for this method is the same as the one in Sect. 2. The considerations on data placement and static resilience of $\mathbf{c}_j$ in the set $\mathcal{N}_j$ of nodes are analogous as well, and an optimized version is obtained similarly as for the forward differential encoding.

3.2   Object Retrieval

Suppose that l versions of a data object have been archived, and the user needs to retrieve the latest version $\mathbf{x}_l$. In the reverse DEC, unlike Subsect. 2.1, the latest version $\mathbf{x}_l$ is encoded as $\mathbf{G}\mathbf{x}_l$. Hence, the user must access a minimum of k nodes from the set $\mathcal{N}_l$ to recover $\mathbf{x}_l$. To retrieve all the l versions $\{\mathbf{x}_1, \mathbf{x}_2, \dots, \mathbf{x}_l\}$, the user accesses the nodes in the sets $\mathcal{N}_1, \mathcal{N}_2, \dots, \mathcal{N}_l$ to retrieve $\mathbf{z}'_2, \mathbf{z}'_3, \dots, \mathbf{z}'_l, \mathbf{x}_l$,

```
1:  procedure ENCODE(X, G, Σ)
2:      FOR 0 ≤ j ≤ L − 1
3:          IF j = 0
4:              return c₁ = Gx₁;
5:          ELSE (This part summarizes Step j + 1 in the text)
6:              c_{j+1} = Gx_{j+1};
7:              Compute z_{j+1} = x_{j+1} − x_j;
8:              Compute γ_{j+1};
9:              IF γ_{j+1} < k/2
10:                 Compress z_{j+1} as z'_{j+1} = Φ_{γ_{j+1}} z_{j+1};
11:                 return c_j = G_{γ_{j+1}} z_{j+1};
12:             END IF
13:         END IF
14:     END FOR
15: end procedure
```

Fig. 6. Encoding procedure for the reverse DEC

respectively. The objects $\mathbf{z}_2, \mathbf{z}_3, \ldots, \mathbf{z}_l$ are recovered from $\mathbf{z}_2', \mathbf{z}_3', \ldots, \mathbf{z}_l'$, respectively through a sparse-reconstruction procedure, and $\mathbf{x}_j$, $1 \leq j \leq l - 1$, is recursively reconstructed as

$$\mathbf{x}_j = \mathbf{x}_l - \left(\sum_{t=j}^{l} \mathbf{z}_t \right).$$

It is clear that a total of $k + \sum_{j=2}^{l} \min(2\gamma_j, k)$ reads are needed for accessing all the l versions and only k reads for the latest version. The performance metrics of the reverse DEC scheme are also summarized in Table 1 (the last column).

Example 4. For the sparsity profile of Example 3, the storage pattern using reverse DEC is $\{\mathbf{z}_2, \mathbf{z}_3, \ldots, \mathbf{z}_{20}, \mathbf{x}_{20}\}$. The I/O read numbers to access $\{\mathbf{z}_2, \ldots, \mathbf{x}_{20}\}$ are $\{6, 10, 6, 10, 10, 10, 10, 4, 4, 6, 10, 6, 10, 6, 10, 8, 4, 6, 10\}$. The total storage size and the I/O reads to recover all the 20 versions are the same as that of the forward differential method. The I/O numbers to recover $\{\mathbf{x}_1, \mathbf{x}_2, \mathbf{x}_3, \ldots, \mathbf{x}_{20}\}$ are $\{156, 150, 144, 134, 124, 114, 104, 94, 84, 80, 76, 70, 60, 54, 44, 38, 28, 20, 16, 10\}$. Note that I/O number to access the latest version (in this case 20th version) is lower than that of the forward differential scheme. For the optimized step, the corresponding I/O numbers are $\{16, 10, 16, 10, 10, 10, 10, 10, 24, 20, 16, 10, 16, 10, 16, 10, 28, 20, 16, 10\}$.

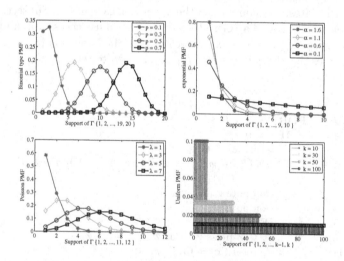

Fig. 7. From top left, clock-wise: Binomial type PMF in p (for k = 20), Truncated exponential PMF in α (for k = 10), Truncated Poisson PMF in λ (for k = 12) and the uniform PMF for different object lengths k. The x-axis of these plots represent the support $\{1, 2, \ldots, k\}$ of the random variable Γ.

3.3   Exploring DEC Benefits with Synthetic Workloads

In this section, we quantify the storage savings offered by the DEC against update patterns that are characterized by in-place alterations. For this study, the update model follows (2), and the in-place alterations are generated from synthetic workloads from a wide-rage of distributions. This exercise showcases the best-case storage savings of DEC as fewer in-place alterations guarantee corresponding sparsity levels in the difference objects, unlike the case of fewer insertions and deletions that totally disturb the sparsity profile.

We present experimental results on the storage size and the number of I/O reads for the different differential encoding schemes. We assume that $\{\Gamma_j, 2 \leq j \leq L\}$ is a set of random variables and its realizations $\{\gamma_j, 2 \leq j \leq L\}$ are known. First we consider a version-control system with $L = 2$, which is the worst-case choice of L as more versions could reveal more storage savings. This setting both (i) serves as a proof of concept, and (ii) already shows the storage savings for this simple case. Later, we also present experimental results for a setup with $L > 2$ versions.

System with $L = 2$ Versions. For $L = 2$, there is one random variable denoted henceforth as Γ, with realization γ. Since Γ is a discrete random variable with finite support, we test the following finite support distributions for our experimental results on the average number of I/O reads for the two versions and the average storage size.

Binomial Type PMF: This is a variation of the standard Binomial distribution given by

$$P_\Gamma(\gamma) = c\frac{k!}{\gamma!(k-\gamma)!}p^\gamma(1-p)^{k-\gamma}, \ \gamma = 1, 2, \ldots, k, \tag{10}$$

where $c = \frac{1}{1-(1-p)^k}$ is the normalizing constant. The change is necessary since $\gamma = 0$ is not a valid event.

Truncated Exponential PMF: This is a finite support version of the exponential distribution in parameter $\alpha > 0$:

$$P_\Gamma(\gamma) = ce^{-\alpha\gamma}. \tag{11}$$

The constant c is chosen such that $\sum_{\gamma=1}^{k} P_\Gamma(\gamma) = 1$.

Truncated Poisson PMF: This is a finite support version of the Poisson distribution in parameter λ given by

$$P_\Gamma(\gamma) = c\frac{\lambda^\gamma e^{-\lambda}}{\gamma!}, \tag{12}$$

where the constant c is chosen such that $\sum_{\gamma=1}^{k} P_\Gamma(\gamma) = 1$

Uniform PMF: This is the standard uniform distribution:

$$P_\Gamma(\gamma) = \frac{1}{k}. \tag{13}$$

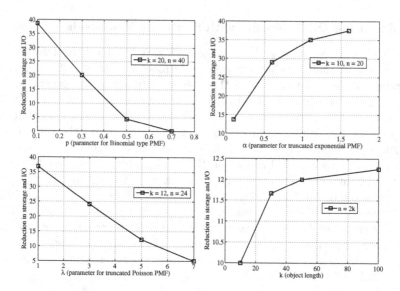

Fig. 8. Average percentage reduction in the I/O reads and storage size for PMFs in Fig. 7 when $L = 2$. The experimental results are presented in the same order as that of the PMFs in Fig. 7.

In Fig. 7, we plot the PMFs in (10), (11), (12) and (13) for various parameters. These PMFs are chosen to represent a wide range of real-world data update scenarios, in the absence of any standard benchmarking dataset (see [16]). The truncated exponential PMFs generate thick concentration for lower sparsity levels, yielding best cases for the differential encodings. The uniform distributions illustrate the benefits of the proposed methods for update patterns with no bias on sparse values. The Binomial distributions provide narrow and bell shaped mass functions concentrated around different sparsity levels. The Poisson PMFs model sparse updates spread over the entire support and concentrated around the center.

For a given PMF $P_\Gamma(\gamma)$, the average storage size for storing the first two versions is $\mathbb{E}[\delta(\mathbf{x}_1, \mathbf{x}_2)] = n + \sum_{\gamma=1}^{k} P_\Gamma(\gamma)\min(2\gamma\kappa, n)$ where $n = \kappa k$. Similarly, the average number of I/O reads to access the first two versions is $\mathbb{E}[\eta(\mathbf{x}_1, \mathbf{x}_2)] = k + \sum_{\gamma=1}^{k} P_\Gamma(\gamma)\min(2\gamma, k)$. When compared to the non-differential method, the average percentage reduction in the I/O reads and the average percentage reduction in the storage size are respectively computed as

$$\frac{2k - \mathbb{E}[\eta(\mathbf{x}_1, \mathbf{x}_2)]}{2k} \times 100 \text{ and } \frac{2n - \mathbb{E}[\delta(\mathbf{x}_1, \mathbf{x}_2)]}{2n} \times 100. \qquad (14)$$

Since $\delta(\mathbf{x}_1, \mathbf{x}_2) = \kappa\eta(\mathbf{x}_1, \mathbf{x}_2)$ and κ is a constant, the numbers in (14) are identical. In Fig. 8, we plot the percentage reduction in the above quantities for the PMFs displayed in Fig. 7. The plots show a significant reduction in the I/O reads (and the storage size) when the distributions are skewed towards smaller γ. However, as expected, the reduction is marginal otherwise. For uniform distribution on Γ, the plot shows that the advantage with the differential technique saturates for large values of k.

We have discussed how the differential technique reduces the storage space at the cost of increased number of I/O reads for the latest version (here the 2nd version) when compared to the non-differential method. For the basic differential encoding, the average number of I/O reads to retrieve the 2nd version is $\mathbb{E}[\eta(\mathbf{x}_2)] = \mathbb{E}[\eta(\mathbf{x}_1, \mathbf{x}_2)]$. However, for the optimized encoding, $\mathbb{E}[\eta(\mathbf{x}_2)] = \sum_{\gamma=1}^{k} \mathrm{P}_\Gamma(\gamma) f(\gamma)$ where $f(\gamma) = k + 2\gamma$ when $\gamma < \frac{k}{2}$, and $f(\gamma) = k$, otherwise. When compared to the non-differential method, we compute the average percentage increase in the I/O reads for retrieving the 2nd version for both the basic and the optimized methods. Numbers for

$$\frac{\mathbb{E}[\eta(\mathbf{x}_2)] - k}{k} \times 100, \tag{15}$$

are shown in Fig. 9, which shows that the optimized method reduces the excess number of I/O reads for the 2nd version.

Experimental Results for $L > 2$. We present the average reduction in the total storage size for a differential system with $L = 10$, assuming identical PMFs on the sparsity levels for every version, i.e., $\mathrm{P}_{\Gamma_j}(\gamma_j) = \mathrm{P}_\Gamma(\gamma)$ for each $2 \leq j \leq 10$. The average percentage reduction in the total storage size and total I/O reads number are computed similarly to (14), and are illustrated in Fig. 10. The plots show further increase in storage savings compared to $L = 2$ case. In reality, the PMFs across different versions may be different and possibly correlated. These results are thus only indicative of the saving magnitude for storing many versions differentially.

To get better insights for $L > 2$, in Fig. 11, we plot the I/O numbers of Examples 3 and 4 for $L = 20$. More than 20% storage space is saved with respect to the non-differential scheme, for only slightly higher I/O for the optimized DEC.

4   Two-Level Differential Erasure Coding

The differential encoding (both forward and the reverse DEC) exploits the sparse nature of the updates to reduce the storage size and the number of I/O reads. Such advantages stem from the application of $\lceil \frac{k}{2} \rceil$ erasure codes matching the different levels of sparsity ($\lceil \frac{k}{2} \rceil - 1$ erasure codes for each $\gamma < \frac{k}{2}$ and one for $\gamma \geq \frac{k}{2}$). If k is large, then the system needs a large number of erasure codes, resulting in an impractical strategy. In this section, we employ only two erasure codes, termed *two-level differential erasure coding*, for the sake of easier implementation, and refer to the earlier differential schemes in Subsects. 2 and 3 as $\frac{k}{2}$-level DEC schemes. We need the following ingredients for the two-level DEC scheme:

(1) An (n, k) erasure code with generator matrix $\mathbf{G} \in \mathbb{F}_q^{n \times k}$ to store the original data object.

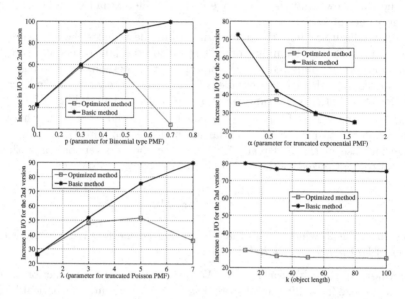

Fig. 9. Average percentage increase in the I/O reads to retrieve the 2nd version for the PMFs (in the same order) in Fig. 7 when $L = 2$. The corresponding values of n and k are same as that of Fig. 8.

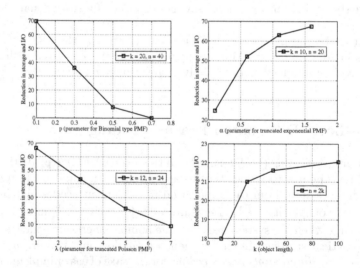

Fig. 10. Average percentage reduction in the I/O reads and total storage size for PMFs in Fig. 7 when $L = 10$. The experimental results are presented in the same order as that of the PMFs in Fig. 7. Identical PMFs are used for the random variable $\{\Gamma_j, 2 \leq j \leq 10\}$ to obtain the results.

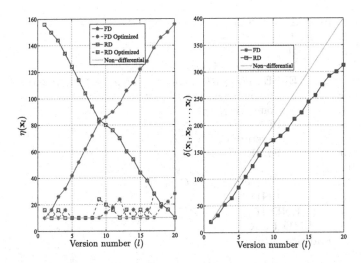

Fig. 11. I/O and storage for Examples 3 and 4. The left plots provide the number of I/O reads to retrieve only the l-th version for $1 \leq l \leq 20$. The right plots show the total storage size till the l-th version for $1 \leq l \leq 20$. Results are for forward and reverse differential methods, with basic and optimized encoding.

(2) A measurement matrix $\Phi_T \in \mathbb{F}_q^{2T \times k}$ to compress sparse updates, where $T \in \{1, 2, \ldots, \lfloor \frac{k}{2} \rfloor\}$ is a chosen threshold.

(3) An $(n_T, 2T)$ erasure code with generator matrix $\mathbf{G}_T \in \mathbb{F}_q^{n_T \times 2T}$ to store the compressed data object. The number n_T is chosen such that $\kappa \triangleq \frac{n}{k} = \frac{n_T}{2T}$.

We discuss only the two-level *forward* DEC scheme. The two-level *reverse* DEC scheme is a straightforward variation.

4.1 Object Encoding

The key point of this encoding is that the number of erasure codes (and the corresponding measurement matrices) to store the γ-sparse vectors for $1 \leq \gamma < \frac{k}{2}$ is reduced from $\lceil \frac{k}{2} \rceil - 1$ to 1. Thus, based on the sparsity level, the update vector is either compressed and then archived, or archived as it. Formally:

Step $j + 1$. Once the version $\mathbf{x}_{j+1}$ is created, using $\mathbf{x}_j$ in the cache, the difference vector $\mathbf{z}_{j+1} = \mathbf{x}_{j+1} - \mathbf{x}_j$ and the corresponding sparsity level γ_{j+1} are computed. If $\gamma_{j+1} > T$, the object $\mathbf{z}_{j+1}$ is encoded as $\mathbf{c}_{j+1} = \mathbf{G}\mathbf{z}_{j+1}$, else $\mathbf{z}_{j+1}$ is first compressed (see (3)) as $\mathbf{z}'_{j+1} = \Phi_T \mathbf{z}_{j+1}$, where the measurement matrix $\Phi_T \in \mathbb{F}_q^{2T \times k}$ is such that any $2T$ of its columns are linearly independent (see Proposition 1). Then, $\mathbf{z}'_{j+1}$ is encoded as $\mathbf{c}_{j+1} = \mathbf{G}_T \mathbf{z}'_{j+1}$, where $\mathbf{G}_T \in \mathbb{F}_q^{n_T \times 2T}$ is the generator matrix of an $(n_T, 2T)$ erasure code. The components of $\mathbf{c}_{j+1}$ are stored across the set $\mathcal{N}_{j+1}$ of nodes.

A summary of the encoding method is provided in Fig. 12.

```
1: procedure ENCODE($\mathcal{X}, \mathbf{G}, \mathbf{G}_T, \Phi_T$)
2:    FOR $0 \leq j \leq L - 1$
3:       IF $j = 0$
4:          return $\mathbf{c}_1 = \mathbf{G}\mathbf{x}_1$;
5:       ELSE (This part summarizes Step $j + 1$ in the text)
6:          Compute $\mathbf{z}_{j+1} = \mathbf{x}_{j+1} - \mathbf{x}_j$;
7:          Compute $\gamma_{j+1}$;
8:          IF $\gamma_{j+1} > T$
9:             return $\mathbf{c}_{j+1} = \mathbf{G}\mathbf{z}_{j+1}$;
10:         ELSE
11:            Compress $\mathbf{z}_{j+1}$ as $\mathbf{z}'_{j+1} = \Phi_T \mathbf{z}_{j+1}$;
12:            return $\mathbf{c}_{j+1} = \mathbf{G}_T \mathbf{z}_{j+1}$;
13:         END IF
14:      END IF
15:   END FOR
16: end procedure
```

Fig. 12. Encoding procedure for two-level DEC

4.2 On the Storage-Overhead

The total storage size for the two-level DEC is $\delta(\mathbf{x}_1, \mathbf{x}_2, \ldots, \mathbf{x}_l) = n + \sum_{j=2}^{l} n_j$, where

$$n_j = \begin{cases} n, & \text{if } \gamma_j > T \\ \kappa 2T, & \text{otherwise.} \end{cases} \tag{16}$$

4.3 Data Retrieval

Similarly to the $\frac{k}{2}$-level DEC scheme, the object $\mathbf{x}_l$ for some $1 \leq l \leq L$ is reconstructed as $\mathbf{x}_l = \mathbf{x}_1 + \sum_{j=2}^{l} \mathbf{z}_j$, by accessing the nodes in the sets $\mathcal{N}_1, \mathcal{N}_2, \ldots, \mathcal{N}_l$. To retrieve $\mathbf{x}_1$, a minimum of k I/O reads is needed. If $\mathbf{z}_j$ is γ_j-sparse and $\gamma_j \leq T$, then $\mathbf{z}'_j$ is first retrieved with $2T$ I/O reads, second, $\mathbf{z}_j$ is decoded from $\mathbf{z}'_j$ and Φ_T through a sparse-reconstruction procedure. On the other hand, if $\gamma_j > T$, then $\mathbf{z}_j$ is recovered with k I/O reads. Overall, the total number of I/O reads for $\mathbf{x}_l$ in the differential set up is $\eta(\mathbf{x}_l) = k + \sum_{j=2}^{l} n_j$, where

$$\eta_j = \begin{cases} 2T, & \text{if } \gamma_j \leq T \\ k, & \text{otherwise.} \end{cases} \tag{17}$$

Similarly, the total number of I/O reads to retrieve the first l versions is also $\eta(\mathbf{x}_1, \ldots, \mathbf{x}_l) = k + \sum_{j=2}^{l} \eta_j$.

Example 5. We apply the threshold $T = 3$ to the sparsity profile in Example 3. The object $\mathbf{z}_{18}$ (with $\gamma_{18} = 4$) is then archived without compression whereas all objects with sparsity lower than or equal to 3 are compressed using a 6×10 measurement matrix. The I/O read numbers to access $\{\mathbf{x}_1, \mathbf{z}_2, \mathbf{z}_3, \ldots, \mathbf{z}_{20}\}$ are

$\{10, 6, 10, 6, 10, 10, 10, 10, 10, 6, 6, 6, 10, 6, 10, 6, 10, 10, 6, 6\}$. The total number of I/O reads to access all the versions is 164 and the corresponding storage size is 328. Thus, with just two levels of compression, the storage-overhead is more than the 5-level DEC scheme but still lower than 400.

4.4 Threshold Design Problem

For the two-level DEC, the total number of I/O reads and the storage size are random variables that are respectively given by $\eta = k + \sum_{j=2}^{L} \eta_j$, where η_j is given in (17) and $\delta = n + \sum_{j=2}^{L} n_j$, where n_j is given in (16). Note that η and δ are also dependent on the threshold T. The threshold T that minimizes the average values of η and δ is given by:

$$T_{\text{opt}} = \arg \min_{T \in \{1, 2, \ldots, \lfloor \frac{k}{2} \rfloor\}} w\mathbb{E}[\delta(\mathbf{x}_1, \mathbf{x}_2)] + (1 - w)\mathbb{E}[\eta(\mathbf{x}_1, \mathbf{x}_2)], \qquad (18)$$

where $0 \leq w \leq 1$ is a parameter that appropriately weighs the importance of storage-overhead and I/O reads overhead, and $\mathbb{E}[\cdot]$ is the expectation operator over the random variables $\{\Gamma_2, \Gamma_3, \ldots, \Gamma_L\}$. This optimization depends on the underlying probability mass functions (PMFs) on $\{\Gamma_j\}$, so we discuss the choice of the parameter $1 \leq T \leq \lfloor \frac{k}{2} \rfloor$ in Subsect. 4.6.

4.5 Cauchy Matrices for Two-Level DEC

Suppose that $\Phi_T \in \mathbb{F}_q^{2T \times k}$ is carved from a Cauchy matrix [22]. A Cauchy matrix is such that any square submatrix is full rank [23]. Thus, there exists a $2\gamma_j \times k$ submatrix $\Phi_T(\mathcal{I}_{2\gamma_j}, :)$ of Φ_T, where $\mathcal{I}_{2\gamma_j} \subset \{1, 2, \ldots, 2T\}$ represents the indices of $2\gamma_j$ rows, for which any $2\gamma_j$ columns are linearly independent, implying that the observations $\mathbf{r} = \Phi_T(\mathcal{I}_{2\gamma_j}, :)\mathbf{z}_j$, can be retrieved from $\mathcal{N}_j$ with $2\gamma_j$ I/O reads. Also, using $\mathbf{r}$ and $\Phi_T(\mathcal{I}_{2\gamma_j}, :)$, the sparse update $\mathbf{z}_j$ can be decoded through a sparse-reconstruction procedure. Thus, the number of I/O reads to get $\mathbf{z}_j$ is reduced from $2T$ to $2\gamma_j$ when $\gamma_j \leq T$. This procedure is applicable for any $\gamma_j < T$. Therefore, a γ_j-sparse vector with $\gamma_j \leq T$ can be recovered with $2\gamma_j$ I/O reads. The total number of I/O reads for $\mathbf{x}_l$ in the two-level DEC with Cauchy matrix is finally $\eta(\mathbf{x}_l) = k + \sum_{j=2}^{l} \eta_j$, where

$$\eta_j = \begin{cases} 2\gamma_j, & \text{if } \gamma_j \leq T \\ k, & \text{otherwise.} \end{cases} \qquad (19)$$

Since the number of I/O reads is potentially different compared to the case without Cauchy matrices, the threshold design problem in (18) can result in different answers for this case. We discuss this optimization problem in Subsect. 4.6.

Example 6. With Cauchy matrix for Φ_T in Example 5, the I/O numbers to access $\{\mathbf{z}_2, \mathbf{z}_3, \ldots, \mathbf{z}_{20}, \mathbf{x}_{20}\}$ are $\{10, 6, 10, 6, 10, 10, 10, 10, 10, 4, 4, 6, 10, 6, 10, 6, 10, 10, 4, 6\}$, which makes the total I/O reads 158. However, the total storage size with Cauchy matrix continues to be 328.

4.6   Code Design Exploration for Two-Level DEC with Synethetic Workloads

In this section, we explore the right choice of threshold T for the two-level DEC scheme. A wide rage of synthetic workloads for the two-level DEC help us identify update patterns where the two-level scheme could be applicable as a substitute to $\frac{k}{2}$-level DEC.

We now present simulation results to choose the threshold parameter $1 \leq T \leq \lfloor \frac{k}{2} \rfloor$ for the two-level DEC scheme in Subsect. 4.4. The optimization problem is given in (18) where

$$\mathbb{E}[\eta(\mathbf{x}_1, \mathbf{x}_2)] = k + \mathrm{Pr}(\gamma \leq T)2T + \mathrm{Pr}(\gamma > T)k,$$

$\mathbb{E}[\delta(\mathbf{x}_1, \mathbf{x}_2)] = \kappa \mathbb{E}[\eta(\mathbf{x}_1, \mathbf{x}_2)]$ and $0 \leq w \leq 1$. Since $\mathbb{E}[\delta(\mathbf{x}_1, \mathbf{x}_2)]$ and $\mathbb{E}[\eta(\mathbf{x}_1, \mathbf{x}_2)]$ are proportional, solving (18) is equivalent to solving instead

$$T_{\mathrm{opt}} = \arg \min_{1 \leq T \leq \lfloor \frac{k}{2} \rfloor} \mathbb{E}[\delta(\mathbf{x}_1, \mathbf{x}_2)]. \tag{20}$$

In Table 2, we list the values of T_{opt}, obtained via exhaustive search over $1 \leq T \leq \lfloor \frac{k}{2} \rfloor$, the average number of I/O reads, the average storage size for the optimized two-level DEC scheme and the $\frac{k}{2}$-level DEC scheme. We denote $\mathbb{E}[\eta(\mathbf{x}_1, \mathbf{x}_2)]$ and $\mathbb{E}[\delta(\mathbf{x}_1, \mathbf{x}_2)]$ by $\mathbb{E}[\eta]$ and $\mathbb{E}[\delta]$, respectively. To compute the average storage size,

Table 2. Optimal threshold value for various PMFs with $k = 10$.

Binomial: $k = 20$, for $\frac{k}{2}$-level: $\eta = 40$ and $\delta = 80$

p	T_{opt}	$\mathbb{E}[\eta]$ (2-level)	$\mathbb{E}[\delta]$ (2-level)	$\mathbb{E}[\eta]$ ($\frac{k}{2}$-level)	$\mathbb{E}[\delta]$ ($\frac{k}{2}$-level)
0.1	3	28.11	56.23	24.55	49.10
0.3	6	35.13	70.27	31.96	63.92
0.5	8	38.99	77.98	38.23	76.47
0.7	9	39.96	79.93	39.95	79.90

Truncated Exponential: $k = 10$, for $\frac{k}{2}$-level: $\eta = 20$ and $\delta = 40$

α	T_{opt}	$\mathbb{E}[\eta]$ (2-level)	$\mathbb{E}[\delta]$ (2-level)	$\mathbb{E}[\eta]$ ($\frac{k}{2}$-level)	$\mathbb{E}[\delta]$ ($\frac{k}{2}$-level)
1.6	1	13.61	27.23	12.50	25.01
1.1	1	14.66	29.32	12.98	25.97
0.6	2	15.79	31.59	14.19	28.39
0.1	2	18.27	36.55	17.26	34.52

Truncated Poisson: $k = 12$, for $\frac{k}{2}$-level: $\eta = 24$ and $\delta = 48$

λ	T_{opt}	$\mathbb{E}[\eta]$ (2-level)	$\mathbb{E}[\delta]$ (2-level)	$\mathbb{E}[\eta]$ ($\frac{k}{2}$-level)	$\mathbb{E}[\delta]$ ($\frac{k}{2}$-level)
1	2	17.01	34.03	15.16	30.32
3	3	20.22	40.45	18.20	36.41
5	4	22.24	44.49	21.06	42.13
7	4	23.29	46.58	22.79	45.58

we use $\kappa = 2$. We see that switching to just two levels of compression incurs negligible loss in the I/O reads (or storage size) when compared to the $\frac{k}{2}$-level DEC scheme. Thus the two-level DEC scheme is a practical solution to reap the benefits of the differential erasure coding strategy.

When Cauchy matrices are used for Φ_T, (18) has to be solved for both

$$\mathbb{E}[\eta(\mathbf{x}_1, \mathbf{x}_2)] = k + \sum_{\gamma=1}^{T} \mathrm{P}_\Gamma(\gamma \leq \gamma)2\gamma + \mathrm{P}_\Gamma(\gamma > T)k$$

$$\mathbb{E}[\delta(\mathbf{x}_1, \mathbf{x}_2)] = n + \mathrm{P}_\Gamma(\gamma \leq T)2T\kappa + \mathrm{P}_\Gamma(\gamma > T)k\kappa.$$

Unlike the non-Cauchy case, $\mathbb{E}[\eta(\mathbf{x}_1, \mathbf{x}_2)]$ and $\mathbb{E}[\delta(\mathbf{x}_1, \mathbf{x}_2)]$ are no more proportional and T_{opt} depends on w, $0 \leq w \leq 1$.

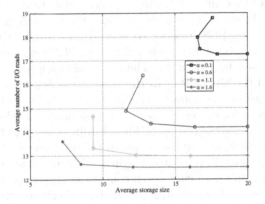

Fig. 13. Average storage size $\mathbb{E}[\delta(\mathbf{x}_1, \mathbf{x}_2)]$ versus average number of I/O reads $\mathbb{E}[\eta(\mathbf{x}_1, \mathbf{x}_2)]$, $1 \leq T \leq \lfloor \frac{k}{2} \rfloor = 5$ with truncated exponential distribution. For each curve, points from left to right tip correspond to $T = \{1, \ldots, \lfloor \frac{k}{2} \rfloor = 5\}$.

To capture the dependency on w, we study the relation between $\mathbb{E}[\eta(\mathbf{x}_1, \mathbf{x}_2)]$ and $\mathbb{E}[\delta(\mathbf{x}_1, \mathbf{x}_2)]$ for $1 \leq T \leq \lfloor \frac{k}{2} \rfloor$. In Fig. 13, we plot

$$\{(\mathbb{E}[\delta(\mathbf{x}_1, \mathbf{x}_2)], \mathbb{E}[\eta(\mathbf{x}_1, \mathbf{x}_2)]), 1 \leq T \leq \frac{k}{2}\}$$

for the exponential PMFs from Subsect. 3.3. For each curve there are $\frac{k}{2} = 5$ points corresponding to $T \in \{1, 2, \ldots, 5\}$ in that sequence from left tip to the right one. The plots indicate the value of $T_{\text{opt}}(w)$ for the two extreme values of w, i.e., $w = 0$ and $w = 1$. We further study the curve corresponding to $\alpha = 0.6$. If minimizing $\mathbb{E}[\eta(\mathbf{x}_1, \mathbf{x}_2)]$ is most important with no constraint on $\mathbb{E}[\delta(\mathbf{x}_1, \mathbf{x}_2)]$ (i.e., $w = 1$), then choose $T_{\text{opt}}(1) = \frac{k}{2}$. This option results in $\mathbb{E}[\eta(\mathbf{x}_1, \mathbf{x}_2)]$ which is as low as for the $\frac{k}{2}$-level DEC scheme. While if minimizing $\mathbb{E}[\delta(\mathbf{x}_1, \mathbf{x}_2)]$ is most important with no constraint on $\mathbb{E}[\eta(\mathbf{x}_1, \mathbf{x}_2)]$ (i.e., $w = 0$), then $T_{\text{opt}}(0) = 2$

results in $\mathbb{E}[\delta(\mathbf{x}_1, \mathbf{x}_2)]$ which is the same as for the 2-level DEC scheme with non-Cauchy matrix. For other values of w, the optimal value depends on whether $w > 0.5$. It can be found via exhaustive search over $1 \leq T \leq \lfloor \frac{k}{2} \rfloor$. In summary, using Cauchy matrix for Φ_T reduces the average number of I/O reads to that of the $\frac{k}{2}$-level DEC with just two levels of compression.

5   Practical Differential Erasure Coding

So far, we developed a theoretical framework for the DEC scheme under a *fixed object length* assumption across successive versions of the data object (see (2)). This assumption typically does not hold in practice because of insertions and deletions, which impact the length of the updated object. In this section, we explain how to control zero pads in the file structure so as to support insertions and deletions in a file, while marginally impacting the storage-overheads.

To exemplify the use of zero pads, consider storing a digital object of size 3781 units through a (12, 8) erasure code of symbol size 500 units, as shown in Fig. 14. Since the object is encoded blockwise, 219 zero pads are added to extend the object size to 4000 units. The zero pads naturally absorb insertions made anywhere in the file, as long as the total size is less than 219 units, thus retaining the length of the updated version to 4000 units. However, since the zero pads are placed at the end, insertions made at the beginning of the file propagate changes across the rest of the file. The difference object is thus unlikely to exhibit sparsity. Alternatively, one could distribute zero pads across the file at different places as shown in the bottom figure of Fig. 14. Here 160 zero pads are distributed at 8 patches with each patch containing 20 zero pads. This strategy arrests propagation of changes when (small size) insertions are made either at the beginning or middle of the file.

Despite zero padding looking like a natural way to handle insertions, it is already clear from this example that the optimization of the size and placements of zero pads is not immediate. We defer this analysis to Sect. 6, and firstly emphasize the functioning of the variable size DEC scheme.

5.1   DEC Step 1 for Variable Size Length Object

Let $\mathcal{F}_1$ be the first version of a file of size V units. The system distributes the file contents into several *chunks*, each of size Δ units. Within each chunk, $\delta < \Delta$ units of zero pads are allocated at the end while the rest of it are dedicated for the file content. Thus, the V units of the file are spread across

$$M = \left\lceil \frac{V}{\Delta - \delta} \right\rceil \tag{21}$$

chunks $\{C_1, C_2, \ldots, C_M\}$, where $\lceil \cdot \rceil$ denotes the ceiling operator. The zero pads added at the end of every chunk promote sparsity in the difference between two successive versions.

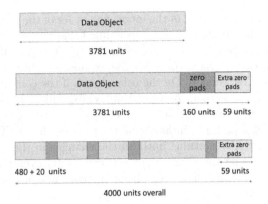

Fig. 14. File structure with different placements of zero pads (ZP) - (i) *ZP-End* where the zero pads are concentrated at the end (middle figure), and (ii) *ZP-Intermediate* where the zero pads are distributed across the file (bottom figure).

Once the file contents are divided into M chunks, they are stored across different servers, using an (n, k) erasure code: the code is applied on a block of k data chunks to output $n(> k)$ chunks which includes the data chunks and $n - k$ encoded chunks that are generated to provide fault tolerance against potential failures. The parameter k is optimized for the architecture with respect to M, which is file dependent:

Case 1: When $M < k$, additional $M - k$ chunks containing zeros are appended to create a block of k chunks. Henceforth, these additional chunks are referred to as *zero chunks*. Then, the k chunks are encoded using an (n, k) erasure code.

Case 2: When $M \geq k$, the M chunks are divided into $G = \lceil \frac{M}{k} \rceil$ groups $\mathcal{G}_1, \mathcal{G}_2, \ldots, \mathcal{G}_G$. The last group $\mathcal{G}_G$ if found short of k chunks is appended with *zero-chunks*. The k chunks in each group are encoded using an (n, k) erasure code.

For the first version $\mathcal{F}_1$, the G groups of chunks together have $\delta M + N\Delta$ units of zero pads, where $1 \leq N < k$, represents the number of zero-chunks added to make $\mathcal{G}_G$ contain k chunks. In addition, the M-th chunk may have extra padding due to the rounding operation in (21). The δM units of zero pads that are distributed across the chunks shield propagation of changes across chunks when an insertion is made in subsequent file versions. This object can now withstand a total of δM units of insertion (anywhere in the file if $\delta M < N\Delta$) by retaining G groups for the second version.

We next discuss the use of zero pads while storing the $(j+1)$-th version $\mathcal{F}_{j+1}$ of the file, $j \geq 1$.

5.2   DEC Step $j + 1$ Under Insertions and Deletions

For the $(j+1)$-th version, the DEC system is designed to identify the difference in the file content size in every chunk. Then the changes in the file contents are carefully updated in the chunks, in the increasing order of the indices $1, 2, \ldots, M$, so as to minimize the number of chunks modified due to changes in one chunk. For $1 \le i \le M$, if the content of C_i grows in size by at most δ units, then some zero pads are removed to make space for the expansion. This C_i will have fewer zero pads than the first version. On the other hand, if the content of C_i grows in size by more than δ units, then the first Δ units of the file content are written to C_i while the remaining units are shifted to C_{i+1}. The existing content of C_{i+1} is in turn shifted, and hence, it will have fewer zero pads than δ. The propagation of changes in the chunks continue until all the changes in the file are reflected. If the insertion size is large enough, then new chunks (or even new groups) have to be added to the existing chunks (or groups), thus changing the object size of the $(j + 1)$-th version.

When file contents are deleted, the zero pads continue to block propagation, this time in the reverse direction. Since deletion results in reduced size of the file contents in chunks, this is equivalent to having additional zero pads (of the same size as that of the deleted patch) in the chunks along with the existing zero pads. After this process, the metadata should reflect the total size of the file contents (potentially less than $\Delta - \delta$) in the modified chunk. Thus, deletion of file contents boosts the capacity of the data structure to shield larger insertions in the next versions.

5.3   Encoding Difference Objects

Note that the differential encoding strategy requires two successive versions to have the same object size to compute the difference. In reverse DEC, once the contents of the $(j + 1)$-th version is updated to the chunks, we compute the difference between the chunks of the j-th and the $(j + 1)$-th version. Then we declare a difference chunk to be non-zero if it contains at least one non-zero element. Within a group, if the number of non-zero chunks, say γ of them, is smaller than $\frac{k}{2}$ then the difference object is compressed to contain 2γ chunks. We continue this procedure of storing the difference objects until the modified object size is at most kG chunks.

A set of consecutive versions of the file that maintains the same number of groups is referred to as a *batch* of versions, while the number of such versions within the batch is called the depth of the batch. The case when insertions change the group size is addressed next as a source for resetting the differential encoding strategy.

5.4   Criteria to Reset DEC

Criterion 1: Starting from the second version, the process of storing the difference objects continues until G remains constant. When the changes require more than G groups, i.e., the updates require more than kG chunks, the system terminates the current batch, and then stores the object in full by redistributing the file contents into a new set of chunks. To illustrate this, let the j-th version of the file (for some $j > 1$) be distributed across M_j chunks, where $\lceil \frac{M_j}{k} \rceil \le G$.

Now, let the changes made to the $(j+1)$-th version occupy M_{j+1} chunks where $\lceil \frac{M_j}{k} \rceil > G$. At this juncture, we reorganize the file contents across several chunks with δ units for zero pads (as done for the first version). After re-initialization, this file has $G' = \lceil \frac{M_{j+1}}{k} \rceil$ groups.

Criterion 2: Another criterion to reset is when the number of non-zero chunks is at least $\frac{k}{2}$ within every group. Due to insufficient sparsity in each group, there would be no saving in storage size in this case, and as a result, a new batch has to be started. However, a key difference from criterion 1 is that the contents of the chunks are not reorganized since the group size has not changed.

6   Experiment Results: Performance of Practical DEC

In this section, we present the performance of the practical DEC technique against a wide spectrum of realistic (but synthetically generated) workloads — that include insertions and deletions, which may lead to change in the overall file size, and so on. We also experiment with real world workloads, specifically, we consider multiple versions of Wikipedia documents to drive our experiments with real data.

Its worth reemphasizing at this juncture, that for these experiments, the update model thus doesn't follow Eq. (2), instead it is as per its practical variant which includes zero pads in the file structure. We showcase experiment results with several workloads capturing wide spectrum of realistic loads to demonstrate the efficacy of our scheme. The main objectives are:

1. to determine the right strategy to place the zero pads in order to promote sufficient sparsity in the difference object for different classes of workloads. This objective is achieved using synthetic workloads of insertions and deletions (see Subsects. 6.1 and 6.2).
2. to present the performance of practical DEC with online datasets such as different versions of Wikipedia pages (see Subsect. 6.3).
3. to compare the storage savings of practical DEC against four baselines, namely (i) a naive technique where each version is fully coded and treated as distinct objects, referred to as non-differential scheme, (ii) selective encoding scheme, a system setup which is fundamentally a delta encoding technique where only the modified chunks are erasure coded and then stored, (iii) Rsync, a well known delta encodng technique for file transfer and synchronization across networks, and finally (iv) gz compresssion, which is applied on individual versions to reduce the storage size (see Subsect. 6.4).

Throughout this section, we use the reverse differential method where the order of storing the difference vectors is reversed as $\{\mathbf{z}_2, \mathbf{z}_3, \ldots, \mathbf{z}_L, \mathbf{x}_L\}$. Also, DEC scheme refers not to the primitive form discussed in Sect. 2, but instead it refers to its variant which was discussed in Sect. 5. Unless specified otherwise, we showcase only the best case storage benefits that come with the application of $\frac{k}{2}$-level DEC scheme, wherein the $\frac{k}{2}$ erasure codes are assumed to have identical storage-overhead of $\kappa = 2$. For the DEC scheme storing two versions, i.e., $L = 2$, the average storage size for the second version is given by

$$\mathbb{E}[\delta(\mathbf{z}_2)] = \kappa\mathbb{E}[\min(2\gamma_j, k)], \qquad (22)$$

which is the average size of the data object after erasure coding. When the storage-overhead κ is held constant for all the $\frac{k}{2}$ erasure codes, we note that the quantity

$$\frac{\mathbb{E}[\delta(\mathbf{z}_2)]}{\kappa} = \mathbb{E}[\min(2\gamma_j, k)], \qquad (23)$$

which is the average storage size prior to erasure coding, is a sufficient statistic to evaluate the placement of zero pads. Henceforth, we use (23) as the yardstick in our analysis. However, in general, when storage-overheads are different, $\mathbb{E}[\delta(\mathbf{z}_2)]$ in (22) is a relevant metric for the analysis. Notice that unlike the quantities in Subsect. 3.3, the quantity in (23) includes raw data as well as zero pads, and this is attributed to a more realistic model of erasure coded versioning system in Sect. 5, where the zero pads facilitate block encoding of arbitrary sized data objects in addition to shielding the rippling effect from insertions and deletions.

6.1 Comparing Different Placements of Zero Pads

We conduct several experiments to compare the storage savings from the zero pads placements highlighted in Fig. 14. The parameters for the experiment are

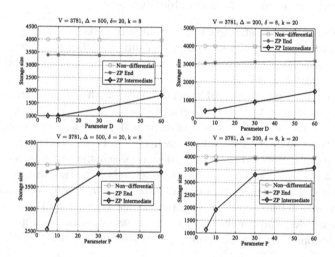

Fig. 15. Comparing different placements of zero pads against insertions: average storage size (as given in (23)) for the 2nd version against workloads comprising random insertions. For the top figures, workloads are bursty insertions whose size is uniformly distributed in the interval $[1, D]$ for $D \in \{5, 10, 30, 60\}$. For the bottom figures, workloads are several single unit insertions whose quantity is distributed uniformly in the interval $[1, P]$, where $P \in \{5, 10, 30, 60\}$.

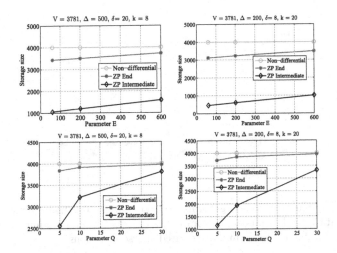

Fig. 16. Comparing different placements of zero pads against deletions: average storage size (as given in (23)) for the second version against workloads comprising random deletions. For the top figures, workloads are single bursty deletions whose size is uniformly distributed in the interval $[1, E]$ for $E \in \{60, 200, 600\}$. For the bottom figures, workloads are several single unit deletions whose quantity is distributed uniformly in the interval $[1, Q]$, where $Q \in \{5, 10, 30\}$.

$V = 3781$, $\Delta = 500, \delta = 20$ and $k = 8$. The two schemes under comparison are *ZP-End* and *ZP-Intermediate* (discussed in Fig. 14), where the zero pads are allocated at the end and at intermediate positions, respectively. Like *ZP-Intermediate* scheme, the *ZP-End* scheme also contains $k = 8$ chunks (each of size Δ), however in this case, 219 zero pads appear at the end in the 8-th chunk. In general, appending zero pads at the end of the data object is a necessity to employ erasure codes of fixed block length. Thus, for the parameters of our experiment, both the *ZP-End* and *ZP-Intermediate* schemes initially have equal number of zero pads (but at different positions), and hence, the comparison is fair.

From our experiments, we compute the average numbers in (23) when two classes of random insertions are made to the first version, namely: (i) single bursty insertion whose size is uniformly distributed in the interval $[1, D]$, for $D = 5, 10, 30, 60$, and (ii) several single unit insertions uniformly distributed across the object, where the number of insertions is uniformly distributed in the interval $[1, P]$, where $P = 5, 10, 30, 60$. We repeat the experiments 1000 times by generating random insertions and then compute the average storage size of the compressed object $\mathbf{z}_2'$ (as given in (23)). In Fig. 15 we plot the average storage size with the *ZP-End* and *ZP-Intermediate* schemes. Similar plots are also presented in Fig. 15 (on the right) with parameters $\Delta = 200, \delta = 20$ and $k = 20$ for the same object. The plots highlight the advantage of distributing the zero pads as it can arrest the propagation of changes through intermediate zero pads. We conduct more experiments for several classes of random deletions

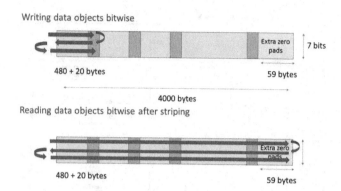

Fig. 17. Bit striping method to generate striped chunks. Top figure depicts bit-level writing of data into the chunks. Bottom figure depicts bit-level reading of data. This technique is suitable for uniformly distributed sparse insertions.

and the results are presented in Fig. 16, which highlight the savings in storage size for the *ZP-Intermediate* scheme.

6.2   Chunks with Bit Striping Strategy

In this section, we analyze the right strategy to synthesize chunks for workloads that involve several single insertions with sufficient spacing. We first explain the motivation for this special case using the following toy example. Consider storing a data object of size $V = 3871$ units using the parameters $\Delta = 500, \delta = 20, k = 8$. Assume that 3 units of insertions are made to the object at the positions $1, 481$ and 961, which translates to modifications of the chunks C_1, C_2 and C_3, respectively. Thus, due to just 3 single unit insertions, three chunks are modified because of which the difference object after compression will be of size 3000 units. Instead, imagine striping every chunk into k partitions at the bit level such that the δ zero pads are equally distributed across the partitions (see the top figure in Fig. 17). Then, create a new set of k chunks as follows: create the t-th chunk for $1 \leq t \leq k$ by concatenating the contents in the t-th partition of all the original chunks (see the bottom figure in Fig. 17). By applying this striping method to the toy example, we see that only one chunk (after striping) is modified, hence, this strategy would need only 1000 units for storage after compression.

For the above example, the insertions are spaced exactly at intra-distance $\Delta - \delta$ units to highlight the benefits, although in practice, the insertions can as well be approximately around that distance to reap the benefits. We conduct experiments by introducing 3 random insertions into the file, where the first position is chosen at random while the second and the third are chosen with intra-distance (with respect to the previous insertion) that is uniformly distributed in the interval $[\Delta - \delta - R, \Delta - \delta + R]$ when $R \in \{40, 80, 120\}$. For this experiment, the average storage size for the second version (i.e., the size of the compressed

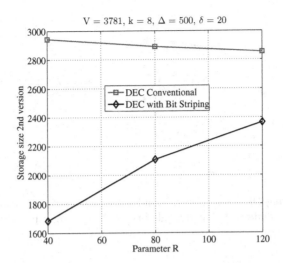

Fig. 18. Comparison of DEC schemes with and without bit striping. Average storage size (given in (23)) for the second version against workload that has 3 single unit insertions with intra-distance uniformly distributed in the interval $[\Delta - \delta - R, \Delta - \delta + R]$, where $R \in \{40, 80, 120\}$. For the experiments, we use $\Delta = 500$ and $\delta = 20$.

object $\mathbf{z}_2'$ given in (23)) is presented in Fig. 18, which shows significant reduction in storage for the striping method when compared to the conventional method. Notice that as R increases, there is higher chance for the neighboring insertions to not fall in the same partition number of different chunks, thus diminishing the gains.

We also test the striping method against two types of workloads, namely, the bursty insertion (with parameter $D \in \{5, 10, 30, 60\}$) and the randomly distributed single insertions with parameter $P \in \{5, 10, 30, 60\}$. For the workloads with single insertions, the spacing between the insertions is uniformly distributed and not necessarily at intra-distance $\Delta - \delta$. In Fig. 19, we present the average storage size for the second version (given in (23)) against such workloads. The plots show significant loss for the striping method against the former workload (as they are not designed for such patterns), whereas the storage savings are approximately close to the conventional method against the latter workload. In summary, if the insertion pattern is known to be distributed a priori, then we advocate the use of the striping method as it provides similar performance as that of the conventional method with a potential to provide reduced storage savings for some special distributed insertions.

6.3 Performance of DEC with Online Datasets

We have conducted experiments based on 5 versions of Wikipedia data on the main article on United States [26], where versions with time stamp "06:21, 15 August 2015" and "00:22, 17th August 2015" are treated as the first

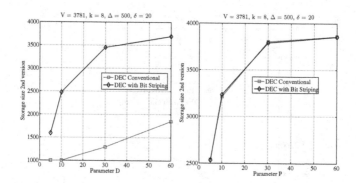

Fig. 19. Comparing DEC schemes with and without bit striping against bursty (the left plot) and randomly distributed single insertions (the right plot) with parameters $D, P \in \{5, 10, 30, 60\}$.

and the fifth versions, respectively. The system parameters for this experiment were

- chunk size (Δ) - 500 bytes
- zero pads in every chunk (δ) - 20 bytes
- number of encoding blocks for erasure coding (k) - 8 chunks

With the above parameters, raw data of 330075 bytes for the first version is expanded to a total of 344000 bytes (by zero pads) and then spread across 86 groups of encoding blocks, where each encoding block contains 8 chunks. For the subsequent versions of the object, changes in different chunks are appropriately identified before storing the difference object as per the DEC scheme. From Version 1 to Version 5, the nature of changes on the chunks are captured in Table 3. For the experiments, we use the reverse differential encoding of Sect. 3 wherein the latest version is encoded in full whereas the preceding versions are stored as difference objects.

Before we proceed to present the storage savings of DEC, the reader may quickly want to know some relevant alternatives for comparison. For cloud storage applications, one straightforward option is to store different versions as

Table 3. Nature of changes at the chunk-level on Wikipedia dataset. The first version has 86 groups of chunks with parameters $\Delta = 500$, $\delta = 20$ and $k = 8$.

Version number	Changes	Affected chunks	Affected groups
1	-	-	-
2	162 bytes removed	5 and 6	1
3	162 bytes added	5 and 6	1
4	102 bytes added	119–124	15 and 16
5	123 bytes added	109–115	14 and 15

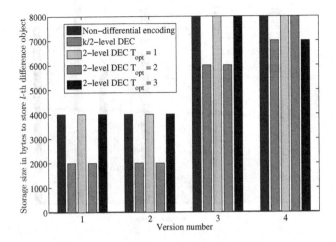

Fig. 20. Performance of DEC on 5 versions of Wikipedia dataset: Storage size (in bytes) needed to store the l-th version, for $1 \leq l \leq 4$ at the end of 5th version, for the following schemes (i) non-differential encoding, (ii) $\frac{k}{2}$-level DEC, (iii) 2-level DEC, with $T_{opt} = 1$(iv) 2-level DEC with $T_{opt} = 2$ and (v) 2-level DEC with $T_{opt} = 3$. Note that the 5th version being the latest is encoded in full, and hence is not presented in the plot.

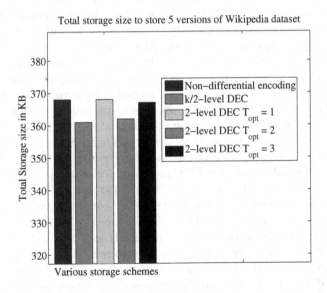

Fig. 21. Performance of DEC on 5 versions of Wikipedia dataset: Total storage size (in KB) needed to store all the 5 versions. The plots indicate that the 2-level DEC scheme with $T_{opt} = 2$ provides storage savings close to the $\frac{k}{2}$-level DEC.

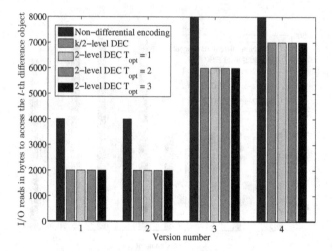

Fig. 22. Performance of DEC on 5 versions of Wikipedia dataset: I/O reads (in bytes) needed to retrieve the l-th version, for $1 \leq l \leq 4$ at the end of 5th version. Although the schemes under comparison provide different storage savings (as shown in Fig. 21), their I/O capabilities are the same due to the use of Cauchy matrices as discussed in Sect. 4.

standalone objects without capitalizing on the *correlation* between successive versions. Since each version is encoded independently, zero pads are needed at the end of the object only to generate the required number of chunks (to be multiple of $k = 8$) for erasure coding. Evidently this method requires around 5 times the size of the first version. Thus, independent encoding of versions along with the insertion of zero pads at the end of the object performs poorly in storage savings, and hence we do not present these numbers. To improve upon this naive scheme, an alternative is to place some zero pads at the end of every encoding group (in the k-th chunk), and then apply non-differential encoding on every modified block. We refer this as the non-differential method in this section.

In Fig. 20, we compare the storage size needed to store the difference objects of the l-th version for $1 \leq l \leq 4$, at the end of 5 versions for the non-differential and the DEC methods. Under the DEC methods, performance of 2-level schemes of Sect. 4 are also presented in Fig. 20, which shows that $T_{\text{opt}} = 2$ provides storage savings close to that of the $\frac{k}{2}$-level DEC scheme. Note that since there are no single-chunk changes, the 2-level DEC scheme with $T_{\text{opt}} = 1$ is as suboptimal as the non-differential scheme. In Fig. 21, we present the total storage size offered by these schemes to store the 5 versions of Wikipedia dataset. Also, in Fig. 22, we present the I/O performance of the DEC schemes in order to retrieve the difference objects. The figure shows that although the 2-level DEC schemes are not efficient in storage size, their I/O performance is as good as the $\frac{k}{2}$-level DEC. This observation is consistent with the theory that erasure codes from Cauchy matrices (discussed in Subsect. 4.5) help reduce I/O reads when retrieving a sparse data object. Overall, Figs. 20, 21 and 22 confirm that DEC

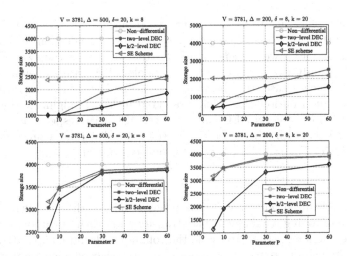

Fig. 23. DEC vs. Selective Encoding with respect to insertions: average storage size for the 2nd version against workloads comprising random insertions. The parameters D and P are as defined for Fig. 15. The $\frac{k}{2}$-level DEC scheme applies an erasure code for each sparsity level, whereas the two-level DEC applies only two erasure codes based on the threshold T_{opt}. The left and the right plots are for bursty and distributed single insertions, respectively.

provides significant reduction in storage size for storing the difference objects without having to look for the modified chunks in every group.

6.4   Comparison with Standard Baselines

Selective Encoding. An important baseline for comparison is a system setup called Selective Encoding (SE), which is a delta encoding technique where only the modified chunks between the successive versions are erasure coded and saved, i.e., visualize the two versions as arrays of numbers, then zero pad the shorter version to make their length equal, and finally store only the changed numbers after component-wise comparison. Observe that the SE scheme is effective if the updated version retains its size, and has few in-place modifications. However, if the object size changes due to insertions or deletions, or when changes propagate at bit level, then dividing the updated version into fixed size chunks need not result in sufficient sparsity in the difference object across versions. For the SE scheme, although there are no preallocated zero pads, they indirectly appear at the end to generate k (or its multiple) number of chunks. We conduct more experiments to compare the storage savings offered by DEC and SE. This time the parameters of the experiment are $V = 3871$, $\Delta = 500, \delta = 20, k = 8$, and the workload includes random insertions with the same parameters as that for Fig. 15. Similar to the preceding experiments, in this section the storage size of the second version includes raw data and zero pads. For the SE based method, zero pads appear at the end to generate $k = 8$ number of chunks from $V = 3871$

units of data. Since, for this experiment, the total number of zero pads is held constant for the two schemes, the comparison is fair. In addition to showcasing the savings of DEC, we present in Fig. 23 the savings of the two-level DEC scheme where only two erasure codes are employed to cater different levels of sparsity. For such a case, the threshold T_{opt} is empirically computed based on the insertion distribution. The plots presented in Fig. 23 highlight the storage savings of both the $\frac{k}{2}$-level DEC and two-level DEC with respect to SE, against bursty insertions (with parameter D). However, for distributed single insertions (with parameter P), only the $\frac{k}{2}$-level DEC outperforms SE, but not the two-level DEC.

Rsync. An advanced version of SE is a storage scheme with concepts from Rsync [19], a delta encoding technique for file transfer and synchronization across networks. The key idea behind Rsync is the rolling checksum computation, using which only the modified/new blocks between successive versions are transferred, thereby reducing the communication bandwidth. With the application of Rsync idea to store versioned data, checksums (or signatures) would have to be computed on every chunk of the j-th version before communicating them to the server containing the $(j+1)$-th version. Subsequently, the server containing latest version rolls over the entire file at fine granularity in search of existing chunks by comparing their checksums, akin to sliding window concept. Finally, the offsets of the found chunks (w.r.t to their position indices in the new file) are returned

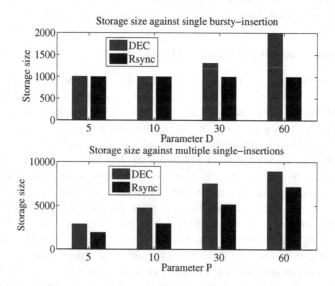

Fig. 24. DEC vs. Rsync with respect to insertions: average storage size for the 2nd version against workloads comprising random insertions. The parameters D and P are as defined for Fig. 15. The plots indicate that Rsync outperfoms DEC in terms of storage savings against both single bursty-insertions and multiple single-insertions.

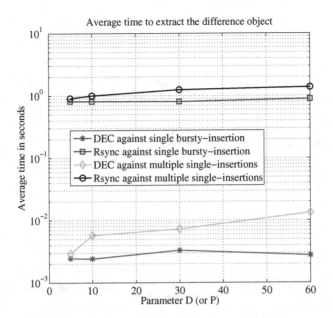

Fig. 25. DEC vs. Rsync with respect to insertions: average time (in seconds) to extract the difference object from the second version in experiments against workloads comprising random insertions. The Rsync scheme consumes substantial time to extract the difference object as it has to apply the sliding-window algorithm at the byte-level in order to look for duplicated chunks in the second version. However, the DEC scheme is computationally efficient due to one-time subtraction of the two versions of data object.

along with the new file contents and corresponding offsets. We note that the rolling checksum computation, which works at unit-level granularity across the file, can be thought as a replacement to the low-complexity subtraction operation in DEC. However, the advantage of reduced computational complexity of DEC comes at the cost of additional storage size for zero pads. Also, since insertions and deletions appear at arbitrary positions in the file, metadata for Rsync should harbor offsets (a.k.a positions of new contents in the file) at unit-level granularity. However, such information is stored at the chunk-level granularity in DEC, thus making it relatively simple in terms of metadata management.

We have conducted experiments to compare the performance of Rsync and DEC. The system parameters for the experiments were $V = 8740$ units, $\Delta = 500$, $\delta = 20$ and $k = 8$. The experiments were conducted against workloads comprising random insertions characterized by single bursty-insertions with parameter $D \in \{5, 10, 30, 60\}$, and multiple single-insertions with parameter $P \in \{5, 10, 30, 60\}$. The average storage size of the difference objects for the two methods are presented in Fig. 24, which shows that Rsync outperforms DEC. This observation can be attributed to the fact that the rolling checksum algorithm of Rsync is powerful enough to search for duplicated chunks in the

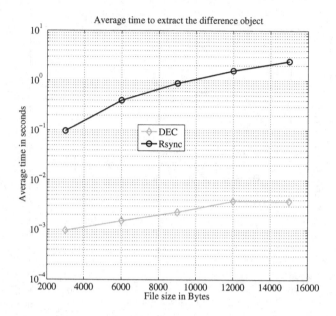

Fig. 26. DEC vs. Rsync with respect to insertions: average time (in seconds) to extract the difference object from the second version in experiments against insertion workloads of different file sizes.

second version, except for those where the insertions were made. Specifically, the initial file of size $V = 8740$ units is first broken into 18 chunks each of size 500 units. Subsequently, each one of the 18 chunks is searched for duplication in the second version by a sliding window comparison that involves 18×8240 chunk-level subtractions in the worst-case. On the other hand, the DEC scheme involves only 24 chunk-level subtractions to obtain the difference object. Thus, the Rsync scheme provides reduced storage size than DEC by trading-off computational complexity. In general, if the file size is V units and the number of chunks is C, then the total number of chunk-level subtractions for Rysnc is of the order $O((V+I)C)$, with I being the insertion size. whereas the corresponding number in DEC is $O(C)$, assuming I is less than that the total number of zero pads. To capture the difference in the computation time between the two schemes, we measure the average time to extract the difference objects against random insertions. The software routines for extracting difference objects (for both Rsync and DEC) were implemented on 64 bit Intel(R) Core(TM) processor @2.13 GHz. The measured average time duration are presented in Figs. 25 and 26, which show a significant difference in the processing time of the two schemes. Overall, we summarize the differences between Rsync and DEC in Table 4. Other than the storage size and computational complexity features, we have also listed erasure coding management as a distinguishing feature. Note that DEC has provision for using just two erasure codes to cater to different sparsity levels, and so, erasure coding management is easy. On the other hand, since Rsync extracts

Table 4. Summary of comparison between Rsync and DEC

Feature	DEC	Rsync
Storage size	Low	Lower than DEC
Computational complexity	Low	High
Erasure coding management	Easy	Complex

only the modified blocks (that are of arbitrary size), it explicitly requires an erasure code for every sparsity level especially when the symbol-size for erasure coding is fixed.

Compression Techniques. One more baseline for comparison is from the set of standard file compression algorithms that are employed to store different versions of a data object. Typically, a file compression algorithm exploits redundancy within the file to generate a compressed file of considerably smaller size. In contrast to such compression schemes, DEC scheme exploits redundancy across versions instead of redundancy within a single version. In order to compare the two schemes, we use the Wikipedia dataset of Subsect. 6.3 to compute the storage savings of the two schemes. We use $\{V_1, V_2, \ldots, V_5\}$ to denote those 5 raw versions. Then, we use the gz compression available online at [25] to first generate the compressed counterparts of the 5 versions of Wikipedia pages, namely $\{W_1, W_2, \ldots, W_5\}$. The file sizes of the 5 versions before and after gz compression are given in the 2nd and 3rd columns of Table 5, respectively, which show that compressed versions are close to 50% of the original size. Subsequently, we compute the sparsity across subsequent versions of the compressed objects, i.e., sparsity of $\{W_1 - W_2, W_2 - W_3, W_3 - W_4, W_4 - W_5\}$, and then compare those

Table 5. Sparsity across successive versions of the Wikipedia dataset in Subsect. 6.3. Compressed versions of the raw data are obtained using [25]. Although the changes on raw data are fewer, compression of individual versions does not promote sparsity across compressed versions. Observe that the % of non-zeros is higher across subsequent compressed versions. Here, low % of non-zero entries implies high sparsity.

Version number	Size of raw data in Bytes	Size of compressed data in Bytes	% of non-zeros across successive versions of raw data of (high sparsity)	% of non-zeros across successive versions compressed data (low sparsity)
1	330593	157068	-	-
2	330429	157060	0.30%	98.41%
3	330593	157068	0.302%	98.41%
4	330701	157096	0.9%	98.21%
5	330824	157148	1.05%	81.45%

values with that obtained from the raw Wikipedia data. For the compressed objects, sparsity values are computed by suitably zero padding the shorter of the two versions. For instance, W_2 is appended with 8 bytes of zero pads to compute the sparsity w.r.t W_1. Although the gz compression algorithm reduces the size of each version, it mixes the contents within each version in such a way that subsequent versions differ at majority of positions despite few in-place alterations on raw data. Thus, % of zeros across subsequent compressed versions are expected to be high, and therefore, applying DEC on the compressed objects may not bring any more reductions in the storage size. On the other hand, we have already shown that DEC enables to store the raw versions at lower storage-size due to high sparsity across raw versions. In Table 5, we present the % of non-zero entries in the difference objects of both raw and compressed data, and the numbers indicate that DEC schemes are effective on raw data than on compressed versions. Thus, the total storage size offered by DEC is substantially smaller than that by individual compression.

7 Concluding Remarks

This paper proposes differential erasure coding techniques for improving storage efficiency and I/O reads while archiving multiple versions of data. Our evaluations demonstrate tremendous savings in storage. Moreover, in comparison to a system storing every version individually, the optimized reverse DEC retains the same I/O performance for reading the latest version (which is most typical), while reducing significantly the I/O overheads when all versions are accessed, in lieu of minor deterioration for fetching specific older versions (an infrequent event). Future works aim at integrating the proposed framework to full-fledged version management systems.

Acknowledgements. This work was supported by the MoE Tier-2 grant "eCode: erasure codes for data center environments" (MOE2013-T2-1-068).

References

1. Huang, C., Simitci, H., Xu, Y., Ogus, A., Calder, B., Gopalan, P., Li, J., Yekhanin, S.: Erasure coding in windows azure storage. In: The Proceedings of the USENIX Annual Technical Conference (ATC) (2012)
2. Thusoo, A., Shao, Z., Anthony, S., Borthakur, D., Jain, N., Sen Sarma, J., Murthy, R., Liu, H.: Data warehousing and analytics infrastructure at Facebook. In: The Proceedings of the 2010 ACM SIGMOD International Conference on Management of Data (2010)
3. Ford, D., Labelle, F., Popovici, F.I., Stokely, M., Truong, V.A., Barroso, L., Grimes, C., Quinlan, S.: Availability in globally distributed storage systems. In: The 9th USENIX Conference on Operating Systems Designand Implementation (OSDI) (2010)
4. Dimakis, A.G., Ramchandran, K., Wu, Y., Suh, C.: A survey on network codes for distributed storage. Proc. IEEE **99**, 476–489 (2011)

5. Oggier, F., Datta, A.: Coding techniques for repairability in networked distributed storage systems. In: Foundations and Trends in Communications and Information Theory, vol. 9, no. 4, pp. 383–466. Now Publishers, June 2013

6. http://subversion.apache.org/

7. http://www.ibm.com/developerworks/tivoli/library/t-snaptsm1/index.html

8. Borthakur, D.: HDFS and Erasure Codes (HDFS-RAID), August 2009. http://hadoopblog.blogspot.com/2009/08/hdfs-and-erasure-codes-hdfs-raid.html

9. The Coding for Distributed Storage wiki. http://storagewiki.ece.utexas.edu/

10. Wang, Z., Cadambe, V.: Multi-version Coding for Distributed Storage. In: Proceedings of IEEE ISIT 2014, Honalulu, USA (2014)

11. Rouayheb, S., Goparaju, S., Kiah, H., Milenkovic, O.: Synchronising edits in distributed storage networks. In: The Proceedings of the IEEE International Symposium on Information Theory, Hong Kong (2015)

12. Rawat, A., Vishwanath, S., Bhowmick, A., Soljanin, E.: Update efficient codes for distributed storage. In: IEEE International Symposium on Information Theory (2011)

13. Han, Y., Pai, H.-T., Zheng, R., Varshney, P.K.: Update-efficient regenerating codes with minimum per-node storage. In: Proceedings of IEEE International Symposium on Information Theory (ISIT 2013), Istanbul (2013)

14. Mazumdar, A., Wornell, G.W., Chandar, V.: Update efficient codes for error correction. In: The Proceedings of IEEE IEEE International Symposium on Information Theory, Cambridge, MA, pp. 1558–1562, July 2012

15. Esmaili, K.S., Chiniah, A., Datta, A.: Efficient updates in cross-object erasure-coded storage systems. In: IEEE International Conference on Big Data, Silicon Valley, CA, October 2013

16. Tarasov, V., Mudrankit, A., Buik, W., Shilane, P., Kuenning, G., Zadok, E.: Generating realistic datasets for deduplication analysis. In: The Proceedings of the 2012 USENIX Conference on Annual Technical Conference (2012)

17. Harshan, J., Oggier, F., Datta, A.: Sparsity exploiting erasure coding for resilient storage and efficient I, O access in delta based versioning systems. In: The Proceedings of IEEE ICDCS, Columbus, Ohio, USA (2015)

18. Harshan, J., Oggier, F., Datta, A.: Sparsity exploiting erasure coding for distributed storage of versioned data. Computing 98, 1305–1329 (2016). Springer

19. http://rsync.samba.org/

20. Donoho, D.L.: Compressed sensing. IEEE Trans. Inf. Theor. 52(4), 1289–1306 (2006)

21. Zhang, F., Pfister, H.D.: Compressed sensing and linear codes over real numbers. In: Information Theory and Applications Workshop (ITA) (2008)

22. McWilliams, F.J., Sloane, N.J.A.: The Theory of Error Correcting Codes. North Holland, Amsterdam (1977)

23. Lacan, J., Fimes, J.: A construction of matrices with no singular square submatrices. In: The Proceedings of International Conference on Finite Fields and Applications, pp. 145–147 (2003)

24. Harshan, J., Datta, A., Oggier, F.: DiVers: an erasure code based storage architecture for versioning exploiting sparsity. Future Gener. Comput. Syst. 59, 47–62 (2016). Elsevier

25. http://www.txtwizard.net/compression

26. https://en.wikipedia.org/wiki/United_States

Secure Integration of Third Party Components in a Model-Driven Approach

Marian Borek$^{(\boxtimes)}$, Kurt Stenzel, Kuzman Katkalov, and Wolfgang Reif

Department of Software Engineering, University of Augsburg, Augsburg, Germany
{borek,stenzel,katkalov,reif}@informatik.uni-augsburg.de

Abstract. Model-driven approaches facilitate the development of applications by introducing domain-specific abstractions. Our model-driven approach called SecureMDD supports the domain of security-critical applications that use web services. Because many applications use external web services (i.e. services developed and provided by someone else), the integration of such web services is an important task of a model-driven approach. In this paper we present an approach to integrate and exchange external developed web services that use standard or non-standard cryptographic protocols, in security-critical applications. All necessary information is defined in an abstract way in the application model, which means that no manual changes of the generated code are necessary. We also show how security properties for the whole system including external web services can be defined and proved. For demonstration we use an electronic ticketing case study that integrates an external payment service.

Keywords: Third party components · Model-driven development · Security-critical systems · Formal specification · UML · SecureMDD · WS-Security

1 Introduction

The use of external web services is essential for many applications. For example, an electronic ticketing system needs to communicate with different external services, such as payment services, currency services or services to deliver the tickets via mail or text message. Therefore, a model-driven approach for developing such applications needs to support the integration of external web services. One way to invoke external web services from a modeled application is to extend the generated code manually. However, our model-driven approach generates from a UML application model runnable code as well as a formal specification for verification of security properties for that application. The manual extension of the generated code would introduce a gap between the formal model and the runnable code, so that the verified properties do not hold necessarily for the running code. Another way is to integrate the external web service into the application model and generate everything from that model. Thereby, everything that is application-specific has to be modeled (e.g., message conversion or

© Springer-Verlag GmbH Germany 2016
A. Hameurlain et al. (Eds.): TLDKS XXX, LNCS 10130, pp. 66–86, 2016.
DOI: 10.1007/978-3-662-54054-1_3

security mechanism) in order to be considered by formal verification. Furthermore, it is often necessary to be able to exchange those services against cheaper, more popular or more efficient ones. The challenge is to make the replacement of external services very easy and minimize verification effort.

Another benefit of the integration of external web services in a model-driven approach is the possibility to extend the approach by application-specific functionality without changing the transformations for code and formal specifications. With this approach also libraries and legacy systems can be integrated by being wrapped inside a web service.

This paper focuses on the integration and exchangeability of external web services in a model-driven approach for security-critical applications by considering different cryptographic mechanisms and describes the verification of the entire application including the communication with external services.

This paper is structured as follows. Section 2 gives an overview of our model-driven approach and Sect. 3 depicts the case study that is used as running example. Section 4 describes the integration and exchangeability of external web services. Section 5 considers the relationship between security properties, assumptions and assurances and describes how the security properties and the assumptions are expressed with regard to the proxy and how an assurance of an external web service can look and how it could be expressed. Section 6 describes the formal model and the verification especially in the context of exchangeability. Section 7 explains how external web services that use cryptography mechanisms can be integrated and exchanged. Section 8 discusses related work and Sect. 9 concludes and discusses some future work.

2 The SecureMDD Approach

SecureMDD is a model-driven approach to develop secure applications. From a UML application model using a predefined UML profile and a platform-independent and domain-specific language called Model Extension Language (MEL [7,20]), runnable code for different platforms as well as formal specifications are generated (see Fig. 1). One formal specification is used for interactive verification with KIV [4] (see [21,22]) and the other to find vulnerabilities with the model-checker platform AVANTSSAR [1] (see [8]). For model-checking the approach supports additional abstractions inside the application model to reduce the complexity and enable model-checking of large applications without manual changes at the generated specifications (see [10]). Additionally, platform-specific models are generated for incremental transformations and better documentation. The approach supports smart cards (implemented in Java Card [26]), user devices like secure terminals or home PCs (implemented in Java), web services (also Java) and external web services. The static view of an application is modeled with UML class diagrams and deployment diagrams. The class diagrams contain the system components, their attributes and the messages that are exchanged between the system components. The deployment diagram contains the communication channels between the system components,

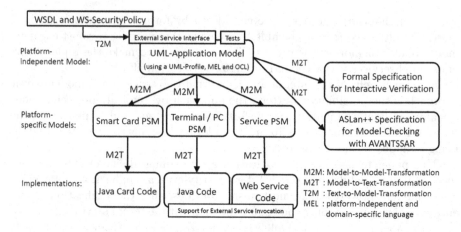

Fig. 1. SecureMDD approach

the attacker abilities and the security mechanisms applied on the communication channels. The dynamic behavior of system components is modeled in UML activity diagrams with our platform-independent and domain-specific language MEL. Application-specific security properties are expressed with Object Constraint Language (OCL) in class diagrams (see [9]) and test cases that generate code for testing the generated application are modeled in UML sequence and activity diagrams (see [17]). The approach is fully tool-supported and all model transformations are implemented. For further information about our approach visit the SecureMDD website[1].

3   eTicket: A Case Study

eTicket is a smart card based electronic ticketing system. Figure 2 shows the system components, the communication channels between the system components and the attacker abilities. The system is designed for an arbitrary number of users and inspectors. Each user owns a smart card (*ETicketCard*) and can buy tickets online with his personal computer (*UserDevice*). A ticket is issued and stored by the *ETicketServer* until the ticket is received and stored by the user's smart card. The inspectors have inspector devices (*InspectorDevice*) to validate and stamp the tickets that are stored on the users' smart cards. Furthermore, the users can manage their tickets that are stored on their smart cards. The attacker abilities are modeled by the stereotype «Threat» with the properties *read, send, suppress*. The stereotype can be applied to any communication channel with different attacker abilities. Hence different attacker behaviors can be modelled. This case study considers a Dolev-Yao attacker [14] who has the abilities to read, send and suppress all messages that are exchanged between

---

[1] www.isse.de/securemdd.

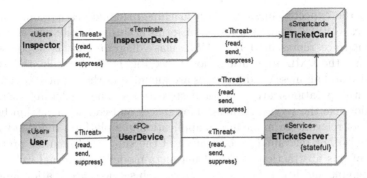

Fig. 2. Deployment diagram of the application

the system components. With regard to the attacker the presented system guarantees three security properties. The first property is "Only tickets that were issued by the eTicket server will be stamped by an inspector" (an inspector can differentiate between real and fake tickets). The basic idea to ensure this property is that only tickets that are issued by the eTicket server can be stored on a valid card (with a private key to a valid certificate) and only tickets that are stored on a valid card can be stamped. The second property is "a ticket will not be lost because of interruptions (e.g. connection errors, an active attacker or a removed smart card) during the buy ticket process". The basic idea is that the ticket is stored on the server until it has been received by the card. The third property is "a ticket will not be used multiple times" and the basic idea is that an issued ticket contains a card id and can only be stored on the belonging card exactly one time. Additionally only tickets that are stored on a valid card can be stamped. This is achieved by security protocols that use cryptographic mechanism like signatures and encryption but also application logic. More detailed information about the security properties and the security mechanism can be found in [8]. To guarantee such security properties they are specified in OCL on the application model and are translated automatically into a formal proof obligation. The complete application model with all diagrams can be found on our website[2]. In the following, the electronic ticketing system is extended by integrating the existing payment service Authorize.Net[3].

4 Modeling Communication with External Web Services

To communicate with a web service, the client which invokes the service needs to know its public interface. For SOAP[4] web services this interface is defined by a WSDL[5] document, which provides the offered web service functionality, and

---

[2] www.isse.de/securemdd.
[3] http://www.authorize.net/.
[4] SOAP: originally an acronym for Simple Object Access Protocol.
[5] WSDL: Web Services Description Language.

especially the expected messages in a machine-readable description. Our approach takes a WSDL document and transforms it automatically into an external
service interface represented by a UML class diagram that is imported as a
module into the UML application model (see Fig. 1). As a result, the external
web service and all message data types are included in the application model as
classes with operations, attributes and stereotypes. The model abstracts from
information like the service address, namespaces and coding algorithm because
they are not relevant for modeling an application in a platform-independent way
and it is also not relevant for verification of the supported security properties.
Because of this abstraction the resulting meta-model for external web services
remains simple and it can be used for other web service specification languages
like WADL[6]. But the omitted information is available in the generated code as
stubs that are generated automatically from the WSDL specification. We use
WSDL2Java from Apache Axis2[7] with JiBX[8] as our stubs-generator to bind
arbitrary class structures on XML documents. That is important because the
data types from external web services differ from the predefined data types in
our approach. The transformation from WSDL to UML is done by hyperModel[9]
that uses generic XML schema documents as input.

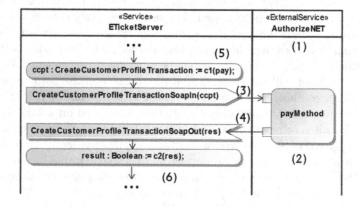

Fig. 3. Invocation of external web service

Figure 3 shows the communication with an external web service. The communication is mainly described in UML activity diagrams using our platform-
independent and domain-specific language called MEL. The external web service
is represented by a UML class with the stereotype «ExternalService» that is
assigned to a UML activity partition (see (1) in Fig. 3) and the external web
method is represented by an operation of that class and a UML call behavior

---

[6] WADL: Web Application Description Language.
[7] axis.apache.org/axis2.
[8] jibx.sourceforge.net.
[9] xmlmodeling.com/hypermodel/.

action without modeling the behavior (2). The invocation of a web method is modeled by a UML send signal action (3) and an accept event action (4) that are connected with the UML call behavior action. For conversion of the message data types between the modeled application and the external service, conversion methods have to be defined. This is done by sub activities using MEL. They have to be invoked before sending a message to an external web service and after receiving a message from the external web service. In more detail Fig. 3 shows how the modeled web service *ETicketServer* invokes the external service *AuthorizeNET* for payment issues. Therefore the conversion methods *c1* (5) and *c2* (6) are used. *c1* converts the data from the modeled application into the required message structure of *AuthorizeNET* and *c2* converts the result of *AuthorizeNET* back. That means the payload *pay* from type *Pay* and *result* from type *Boolean* belong to the modeled application and *CreateCustomerProfileTransactionSoapIn* and *CreateCustomerProfileTransactionSoapOut* are data types used by *AuthorizeNET*.

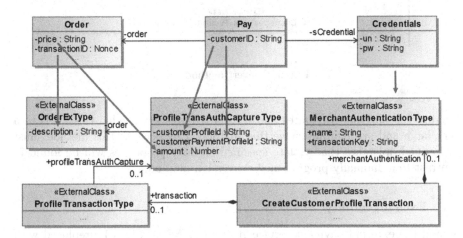

Fig. 4. Converted classes

Figure 5 shows the definition of the conversion method *c1* that converts the *pay* object into *CreateCustomerProfileTransaction* and Fig. 4 illustrates the mapping between the converted classes. Some classes can be mapped one-to-one (e.g., *Credentials* and *MerchantAuthenticationType*). Other classes consist of merged information from different classes (e.g., *ProfileTransAuthCaptureType* contains attributes from the *Pay* and *Order* classes). And if the output class has more attributes than the input class, the missing information has to emerge from the existing one by duplication or transformation (e.g., *ProfileTransAuthCaptureType* needs a *customerProfileId* and a *customerPaymentProfileId* that can be both extracted from *customerID*). Therefore, some generic and predefined methods are used (e.g., substring for string manipulation). The conversion in Fig. 5 is minimal but the output class *CreateCustomerProfileTransaction* has

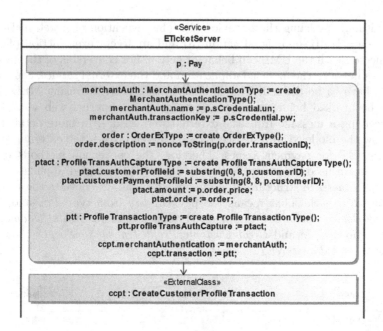

Fig. 5. Convert method *c1*

roughly 100 optional attributes and if all attributes are needed this leads to a very large and error-prone converting method. But because this method must be verified, mistakes that violate specified security properties will be found in contrast to a manually programmed conversion method.

4.1 Exchangeability

Should the external payment service *AuthorizeNET* in Fig. 3 be replaced with a different one, the protocol diagrams have to be changed and the verification of the entire application would need to be redone. To avoid this, the security-critical protocols and the invocation of an external service can be separated. In order to achieve that, we support a proxy pattern (see Fig. 6). Therefore, a proxy interface that is independent from the external payment service has to be modeled and used in the protocols described by activity diagrams. For each external service a proxy has to be modeled that implements this interface and invokes the external service. Then the external web services can be easily switched by changing the proxy in the class diagram. In our case study the *ETicketServer* has to invoke a *PayService* proxy interface. To add a concrete payment service like *AuthorizeNET*, a new proxy (e.g., *AuthorizeNETProxy*) has to be created that inherits from *PayService* and defines the behavior of the *pay* method and the conversion methods.

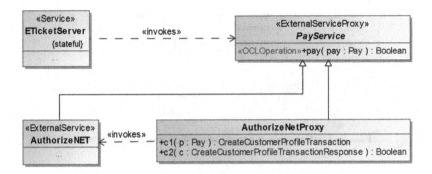

Fig. 6. Proxy pattern for exchangeability of AuthorizeNET

5   Security Properties, Assumptions and Assurances

For the verification of certain security properties of the entire application, assumptions about the external service are necessary. An example for a security property for our electronic ticketing system is that "Only paid tickets will be stamped by an inspector". Obviously, some information about the external payment service method are necessary, e.g., that "if the return value is positive, then the payment was or will be successful". This assumption is specified for the proxy, which represents an abstraction of the external service and manages the conversion between different messages and the invocation of the external service. As a result, the security property is provable independently from the external web service.

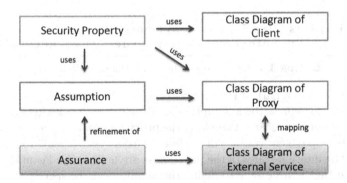

Fig. 7. Relation between security property, assumption and class diagrams

Figure 7 shows that a security property uses classes from the client and the proxy and of course the assumptions specified for the proxy. The assurance of the external service uses classes from the external service interface that is generated from the WSDL and the assurance has to be a refinement of the assumption.

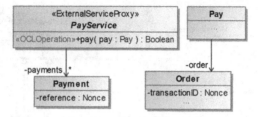

Fig. 8. Partial class diagram of an *ExternalServiceProxy*

The security property for the application, the assurance of the external service and the assumption about the proxy are formally defined with OCL on classes that represent internal states and messages of the application participants. The mapping between messages is handled by the modeled conversion methods that are automatically transformed to executable Java code, and also to formal specifications. The relationship between the internal states is only necessary for verification. Hence, it is not modeled but specified during the verification. Those two mappings make it possible to show the refinement between the assurance of the external service and the assumption defined on the proxy. If the assurance is a refinement of the assumption (the formal definition is shown in Subsect. 6.2), the security property that holds for the payment method of the proxy holds also for the payment method of the external service. This way, the external services can be exchanged without influence on the security property if the assurances of the new external service are also a refinement of the assumptions of the proxy.

5.1   Assumption

```
result   implies   PayService . payments->
    exists ( p |   p . reference   =   pay . order . transactionID )
```

Listing 1.1. OCL postcondition on the operation pay

Figure 8 depicts a part of the proxy's class diagram from the electronic ticketing system. To express the assumption for the electronic ticketing system some internal states are necessary. Therefore, the proxy interface *PayService* gets a list of payments. A payment consists of a reference from type Nonce[10] to identify the payment. The internal states of the proxy class are minimal and contain only the information that is necessary to define the assumption. To specify the assumption "if the return value is positive, then the payment was or will be successful" it is important to abstract the payment process. Each ticket that is issued by the *eTicket* service contains a *ticketID*. This *ticketID* is stored as *transactionID* in the pay object that is received by the pay method as input parameter. The assumption is described as OCL postcondition on the pay method of the proxy

---

[10] A nonce describes an arbitrary and unique number.

interface and specifies that "if the result is true then the list of payments contains a payment with the *transactionID* as reference" (see Listing 1.1).

5.2   Security Property

The security properties are expressed as OCL invariants. The property "Only paid tickets will be stamped by an inspector" is expressible by using the payments from the proxy and the tickets that have been successfully stamped and stored by the inspector devices. Hence, the property is equivalent to the OCL constraint in Listing 1.2. It describes that "the list of payments has to include each stamped ticket".

```
InspectorDevice.allInstances().stampedTicketIDs->
     forAll(id | PayService().payments.
          reference->includes(id) )
```

Listing 1.2. Security Property as OCL Invariant

Because the system is defined for an arbitrary number of inspectors and each inspector has an inspector device to stamp tickets, the OCL expression collects all stamped tickets from all inspector devices.

5.3   Assurance

The assurance of the existing payment service is guaranteed by the service provider. So the assurance is independent from the developed application. That means it uses different data structures, different names and different content.

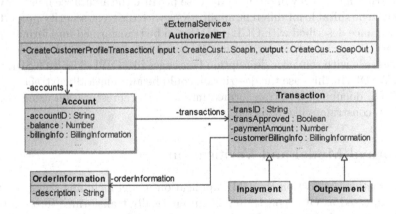

Fig. 9. Possible class diagram of AuthorizeNET

```
(output ... resultCode = MessageTypeEnum.Ok) implies
AuthorizeNET.accounts->exists(a|
     a.accountID = input ... customerProfileId and
     a.auth = input ... merchantAuth and
     a.balance = a.balance@pre+input ... amount and
     a.transactions->exists(t|
          t.oclIsKindOf(Inpayment) and
          input ... amount = a.transactions.amount and
          input ... description = t.orderInfo.description))
```

Listing 1.3. Exemplary assurance as OCL postcondition

Listing 1.3 shows a possible assurance of the payment method of Authorize.Net in OCL and Fig. 9 depicts the respective class diagram. The proxy from Fig. 8 stores only the transaction IDs as *reference*. In contrast the external payment service *AuthorizeNET* in Fig. 9 manages whole accounts for several customers. It stores the *balance*, the authentication data (*auth*) and the transactions including the transaction *amount*, the order information (*orderInfo*) and the transaction type described by the sub classes *Inpayment* and *Outpayment*. The assurance in Listing 1.3 specifies, "the result code *Ok* implies that there exists an account containing the following information":

- "the received profile id"
- "the received authentication data"
- "a balance increased by the received amount"
- "an inpayment transaction with the received amount and description"

For verification some additional information would be necessary (e.g., that the transaction is new or if the payment failed the transactions are not modified). But Listing 1.3 demonstrates that the assurance is more precise than the assumption (Listing 1.1). This will often be the case because the assurance must be very detailed to be suitable for different applications with different security properties. The assurance described with OCL in UML can be transformed into formal specifications that are used for verification. A useful extension would be to describe the assurance with semantic description languages like OWL-S or SAWSDL that enrich WSDL. In this case the description could be automatically extracted from the WSDL document and transformed into a UML representation using classes and OCL constraints like depicted in Listing 1.3.

6   Formal Model and Verification

For interactive verification the UML application model including security properties expressed as OCL constraints is automatically transformed into a formal model with proof obligation. The formal model is based on algebraic specifications and Abstract State Machines (ASMs) [11]. It specifies a world in which system participants (agents) exchange messages according to the protocols, and an attacker tries to break the security. Agents are either users, smart cards, terminals, or services, and there exists an arbitrary (finite) number of agents.

One step of the ASM corresponds to one protocol step (receiving and processing a message), a run of the ASM creates a trace, i.e. a sequence of steps, that describes one possibility of what can happen in this world. Since the ASM is indeterministic it models not one but many traces. The idea is that the ASM models everything that can happen in the real world (with respect to the application). So if the application can be proved secure in the formal model it should be secure in the real world. More details can be found in [6]. The proof obligations are predicate logic formulas contained in algebraic specifications. These formulas are considered to be invariants and therefore checked initially and after each step of the ASM. The transformation of the security properties defined as OCL constraints into algebraic specifications can be found in [9].

6.1 Transformation of Assumption and Conversion Methods

To verify security properties for a system that uses external web services, the assumptions described as OCL pre- and postconditions on the external service methods have to be considered as well. Therefore, the assumptions are transformed automatically into predicate logic formulas and after an external web service method is executed the respective assumption holds. Hence, the behavior of the external service methods that are used inside the protocols is specified formally and the security properties can be verified.

```
1   functions
2         collect47:  (agent  -> listofPayment),
3                             agents -> listofPayment;
4   predicates
5         exists48:  listofPayment ,  Pay;
6
7   procedures
8         pay:  Pay,  agent:  bool,
9               (agent  -> listofPayment);
10
11  axioms
12  collect47_emp:  |-  collect47 (PayService-payments ,[]) = [];
13  collect47_rec:  |-  collect47 (PayService-payments ,ag  ' + agents) =
14                        PayService-payments(ag) +
15                        collect47 (PayService-payments ,agents);
16
17  exists48_emp:  |-  not  exists48 ([] ,a_Pay);
18  exists48_rec:  |-  exists48 (a_Payment  ' + a_listofPayment ,  a_Pay) <->
19        a_Payment . reference_Nonce =
20                        a_Pay . order_Order . transactionID_Nonce and
21        a_Payment . price_string = a_Pay . order_Order . price_string or
22        exists48 (a_listofPayment ,  a_Pay);
23
24  pay:  exPayService(ag) |-
25        <pay(a_Pay ,ag; result -bool-var ,  PayService-payments)>
26        (result -bool-var ->
27                        exists48 ( collect47 ( PayService-payments ,ag) ,a_Pay));
```

Listing 1.4. Assumption as algebraic specification

Listing 1.4 shows the assumption from Fig. 8 translated into an algebraic specification. The assumption is defined on the *pay* operation of the proxy class *PayService*. It is declared in line (8–9) as a procedure. In contrast to a security property that is specified as a predicate, a procedure can be executed as a program. This is necessary because the *pay* operation has to be invoked as part of the dynamic behavior of a modeled system component. The pay procedure extends the signature of the *pay* operation with an agent (in this case *PayService*) and a dynamic function that maps an agent to the value of one of its attributes. In this way the attributes of the agent can be modified. The axiomatisation of the procedure is specified in line (24–27). The procedure pay can be executed as a program if agent *ag* is a *PayService*. After pay has executed, the postcondition in line (26–27) holds. The postcondition is very similar to the OCL postcondition but the OCL operations have to be expressed in predicate logic. For simplicity and because the specifications are generated automatically, for each invocation of an OCL operation a new function or predicate that also specifies the body of the invoked OCL operation is generated. Line (2) declares and line (12–15) define a collect operation that collects all payments from *PayService*. Line (5) and line (17–22) specify the exists query from Listing 1.1 that checks if a payment exists which contains the transaction id from the input message *pay*.

In contrast, the conversion method *c1* from Fig. 5 is not specified in OCL but in UML sub activities with our platform-independent and domain-specific language called MEL. Hence, it is not specified in a declarative way and therefore it is translated into a program that is executed step by step with dynamic logic. The dynamic behavior of the components is modeled and translated in the same way.

6.2   Refinement

For exchangeability, the assumption is independent from the assurance of the real web service. To prove that the security properties also hold for the real web service a refinement between the assumption and the assurance is necessary. This is done by a data refinement and considers only the messages and internal states of the proxy and the external web service. Hence, it is independent from the main application behavior and the security protocols.

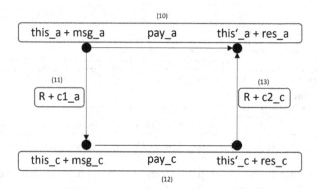

Fig. 10. Data refinement

```
1   pay_a: PayService x Pay -> PayService x Boolean;
2   pay_c: AuthorizeNET x CreateCustomerProfileTransactionSoapIn
3          -> AuthorizeNET x CreateCustomerProfileTransactionSoapOut;
4   R : PayService x AuthorizeNET;
5
6   R(ps, an) <-> forall s. s in ps.payments.reference.nonce <->
7   s in an.accounts.transactions[Inpayment].orderInfo.description;
8
9   ProofObligation:
10  pay_a(this_a, msg_a) = [this'_a, res_a] and
11  R(this_a, this_c) and msg_c = c1_a(msg_a) and
12  pay_c(this_c, msg_c) = [this'_c, res_c]
13  -> c2_c(res_c) = res_a and R(this'_a, this'_c);
```

Listing 1.5. Data Refinement

Figure 10 and Listing 1.5 depict such a data refinement for the electronic ticketing system using the payment service AuthorizeNET. For better comprehension Listing 1.5 uses a pseudo language. *pay_a* describes the operation *pay* from *PayService* with the assumption defined in Listing 1.1 and is formalized as algebraic specification in Listing 1.4. Because the operation modifies the state of *PayService* the procedure receives not only the message *msg_a* from type *Pay*, but also the *PayService* instance *this_a* as input parameters and returns the modified *PayService* instance *this'_a* and the result of the *pay* operation *res_a* of type boolean (see line (1, 10) in Listing 1.5 and (10) in Fig. 10). *pay_c* (see line (2, 12)) is also a procedure and results from the assurance defined in Listing 1.3. The data refinement shows that the concrete *pay_c* is a refinement of the abstract *pay_a*. Therefore, *msg_a* has to be translated into *msg_c* and the result *res_c* has to be translated into *res_a*. This is done by the modeled converting methods c1 and c2 (see Fig. 5) that are translated into the formal procedures *c1_a* and *c2_c*. The conversion between the internal states of *PayService* and *AuthorizeNET* have to be specified by the relation *R* (see line (4–7)). In the considered case study this relationship is very simple because the *PayService* proxy has only a list of payments. The relation defined as predicate describes that each element in *PayService payments* is contained in the inpayment transactions of *AuthorizeNET* and vice versa. To show the refinement, the proof obligation (see line (10–13) in Listing 1.5) has to be proved. In short, if *pay_c* is invoked with the converted input from *pay_a* (see line (11–13)), its converted output is equal to the output of *pay_a* (see line (10, 13)). Because of the different data structures, different names and different information content between the proxy and the external web service, it can be difficult to see that the assurance is a refinement of the assumption. Therefore, the proof obligation has to be proved.

7   Cryptography and External Web Services

Depending on the application's security properties, it is important to model the security mechanisms that are applied on the communication between the client and the external web services. This way, the security mechanisms can be transformed into the formal model and are considered during the verification. There are different ways to secure the communication by cryptography for web services.

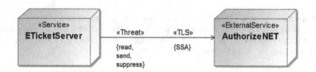

Fig. 11. TLS applied between two system participants

The simplest and most common way is to use Transport Layer Security (TLS [13]). It is a standard protocol that is independent from any specific web service. But TLS does not fulfill all possible requirements, e.g., end-to-end encryption. WS-SecurityPolicy is a language to describe individual cryptographic protocols for web services. But the design of application-specific security protocols is error-prone and requires verification. Additionally, it is likely that different web services have different WS-SecurityPolicies. This influences the exchangeability of web services. Our approach supports three different ways to secure the modeled functionality using cryptography.

1. The first one is to apply TLS on a connection between two system participants, e.g., between the electronic ticketing server and an external payment
   service (see Fig. 11). This is modeled using a stereotype that is applied on a
   UML Communication Path between two UML Nodes in a deployment diagram. Furthermore, the stereotype has two properties to distinguish between
   mutual authentication (MA) and server side authentication (SSA). This security mechanism restricts the attacker's abilities that are modeled with the
   stereotype ≪Threat≫ and its properties *(read, send, suppress)*. In the presented case study the payment service supports SSA and it is invoked over
   an insecure network where an attacker can read, send and suppress messages.
   From this model, runnable code that uses TLS to secure the communication
   as well as the key stores and default keys that have to be exchanged during
   deployment are generated automatically. Additionally, the attacker abilities
   in the formal model are changed so that the attacker can only abort an existing TLS connection or establish a new TLS connection.

2. The second way uses predefined security data types for encryption, signatures, macs, hashes, nonces, keys and predefined operations to create those
   data types. These security mechanisms are used to model the security protocols in the presented case study between the modeled system participants.
   But because in the past our focus was not exchangeability but ensuring
   application-specific security properties, there is no strict separation between
   application logic and cryptography. Hence, for exchangeability this approach
   is unsuitable.

3. Therefore, the third way to secure the modeled functionality using cryptography in our approach is WS-SecurityPolicy. It applies cryptography
   directly before sending a message and directly after receiving a message but

always independent from application logic. WS-SecurityPolicy is integrated in WSDL so the policies can be automatically extracted from the WSDL and transformed to an abstracted UML representation using stereotypes, classes and attributes. Additionally, the UML representation abstracts from WS-SecurityPolicy assertions like *AlgorithmSuite* because they are not used for the formal verification. A WS-SecurityPolicy specification of a web service can contain several alternative policies so the application designer has to choose one that should be used by the client. This is modeled with an attribute of the client or the proxy class.

Because reusability makes software more clear, maintainable and reduces errors, we mapped WS-SecurityPolicies to the already supported notation that is used for the generation of formal specifications. Therefore, MEL expressions whose behavior is equivalent to the policies are injected inside the modeled activity diagram that invokes the external service. This is done with model-to-model transformations in QVTo [24] and the resulting model is used with our existing generator for formal specifications.

Figures 12 and 13 show part of a protocol with injected policy behavior. In the original protocol, without the injected policy, the client collects the payment information (first activity node in Fig. 12) and sends it to the service proxy that invokes the pay method (last activity node in Fig. 13), which handles the conversion and invokes the external service. The regarded WS-SecurityPolicy describes a simple security protocol with symmetric binding and body encryption. The symmetric binding uses a X.509 certificate as protection token that is already exchanged and will be addressed in messages by its thumbprint reference. Hence, the injected part in Fig. 12 (second activity node) generates a symmetric key, stores it in the key store to be able to decrypt an optional response, encrypts the symmetric key with the public key from the X.509 certificate that belongs to the external service, creates the SOAP header including the encrypted symmetric key, encrypts the payment information with the symmetric key and puts it in a SOAP body object. The injected part in Fig. 13 (second activity node) decrypts the symmetric key from the header and uses the symmetric key to decrypt the payment information, that is used to invoke the pay method. The send and receive nodes are modified because the original modeled messages were exchanged with SOAP messages by the transformations. This is all done automatically together with the generation of the required classes. Besides the encrypted symmetric key the real SOAP header contains also algorithmic information that can be omitted and token references like the thumbprint of the public key that is not necessary if the formal representation of the external service has only one key pair.

Because the policy behavior is injected before the pay method (which handles the conversion and invokes the external service) is invoked, replacing an external service that uses WS-SecurityPolicy with a new external service that also uses WS-SecurityPolicy can be done without additional verification if the policy of the replaced external service is a subset of the new one. In this case the new external service ensures the same security properties like

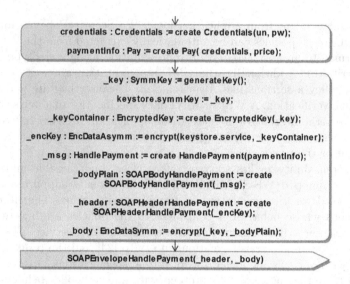

Fig. 12. Send HandlePayment with injected policy behavior

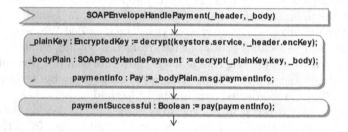

Fig. 13. Receive HandlePayment with injected policy behavior

the old external service plus some additional ones. This can be checked very fast and automatically during the transformation with QVTo.

8   Related Work

There are many works that consider web services in a model-driven approach. The most related works can be mainly categorized in static web service representation, orchestration and security.

Castro et al. [12] describe web services with a UML meta-model that is very close to the WSDL one. That means that each WSDL element is described by a stereotype with the same name. They also define transformation rules that generate a UML model from a WSDL specification. The advantage of this approach is that the WSDL specification and the generated UML model have the same information content, but the disadvantage is that the resulted UML model has the same complexity like the WSDL specification. But the aim of a UML model is abstraction. [15,27] describe approaches that transform a WSDL specification in an abstract UML model by predefined rules. This is very close to our WSDL to UML transformation, but because our abstraction level differs from theirs, we defined modified abstraction rules that fit better to our approach. For example we do not abstract from the messages defined in the WSDL specification because they are used to model the communication with the external web service. They do not describe a model-driven approach that integrates external web services.

Self-Serv [3,5] is a model-driven approach for web service development with focus on orchestration. It uses state machines and generates BPEL[11] based service-skeletons with orchestration logic but without modeling or generating application-specific logic. It does not consider security aspects and it does not integrate existing services. MDD4SOA [18] also illustrates a model-driven approach for web service orchestration. From a UML model code is generated for BPEL, WSDL, Java and the formal language Jolie. Security and the integration of external web services are not considered.

Nakamura et al. [23] enable the model-driven development of WS-Security-Policy specifications. They describe standard security properties with stereotypes that are used to select predefined security patterns from a library and apply them on the model. From that model, configuration documents for IBM WebSphere Application Server (WAS) and WS-SecurityPolicy specifications are generated. In contrast we generate an abstract model from the WS-SecurityPolicy of an existing web service that is selected and applied on the client as well as used for formal verification. Menzel et al. [19] also introduce a model-driven approach that uses abstract security patterns to generate XML-based configuration documents for the Apache Rampart-Modul that implements the WS-Security Stack. Jensen et al. [16] is also a model-driven approach that generates WS-BPEL, WSDL and WS-SecurityPolicy specifications from a model. The mentioned works do not integrate external web services and do not transform WS-SecurityPolicies into a formal representation for verification.

Pironti et al. [25] generate verified client-code, which uses an existing TLS-Service but they have to write thousands of lines of code for the data conversion manually and without security guarantees. In [2] they explore the verification of systems with external services but they do not generate code, and verifying the security of the application is not part of that work.

We are not aware of a model-driven approach that considers the secure integration and replacement of existing web services in security-critical applications and verifies security properties about the whole application including the external services.

9   Conclusion and Future Work

The integration and replacement of external security-critical web services in a model-driven approach is a novel and important topic. It enables the

---

[11] BPEL: WS-Business Process Execution Language.

model-driven development of realistic applications that use existing code, e.g., services, legacy systems or libraries. In this paper our contribution can be summarized as follows. We have shown how to model the communication with external web services and how runnable code is generated automatically from the model without the necessity of manual changes. We have also discussed how security properties for the modeled application that uses external web services can be verified. An important issue was the replacement of external web services with minimal verification effort. Therefore a proxy with assumptions about the behavior of the external web service has to be specified on the application model, translated into the formal model and used to prove that the assurance of the external web service is a refinement of the assumption. Finally we have shown how web services that use different cryptographic protocols are handled. The novelty of the presented work is the modeling of the security properties and the assumptions with regard to a proxy pattern as well as the transformation of the modeled assumption into a formal specification and the refinement between this assumption and the assurance of the external web service. As a result, we were able to develop an electronic ticketing system with our model-driven approach that integrates the real payment service Authorize.Net. Additionally, we are now able to extend our approach by application-specific functionality without changing the transformations for code and formal specifications. This can be done by providing the functionality as a web service and specifying the behavior with OCL. In our opinion this work is a significant extension of model-driven development that includes verification and makes the development of real applications that use external components feasible.

Future work is to express a detailed and complete assurance of an external web service inside WSDL (e.g., with OWL-S[12] or SAWSDL[13]) and transform it automatically into UML and OCL. This way, the web service describes its full behavior in a machine-readable way with formal guarantees that are necessary for the automatic integration of external web services in security critical applications. Furthermore, our approach supports external web services only in combination with interactive verification. It would be interesting if the integration and especially the refinement can be proved automatically with a model-checker.

References

1. Armando, A., et al.: The AVANTSSAR platform for the automated validation of trust and security of service-oriented architectures. In: Flanagan, C., König, B. (eds.) TACAS 2012. LNCS, vol. 7214, pp. 267–282. Springer, Heidelberg (2012). doi:10.1007/978-3-642-28756-5_19
2. Bagheri Hariri, B., Calvanese, D., De Giacomo, G., Deutsch, A., Montali, M.: Verification of relational data-centric dynamic systems with external services. In: Proceedings of the 32nd Symposium on Principles of Database Systems, pp. 163–174. ACM (2013)

---

[12] OWL-S: Semantic Markup for Web Services.
[13] SAWSDL: Semantic Annotations for WSDL and XML Schema.

3. Baïna, K., Benatallah, B., Casati, F., Toumani, F.: Model-Driven web service development. In: Persson, A., Stirna, J. (eds.) CAiSE 2004. LNCS, vol. 3084, pp. 527–543. Springer, Heidelberg (2004). doi:10.1007/978-3-540-25975-6_22
4. Balser, M., Reif, W., Schellhorn, G., Stenzel, K., Thums, A.: Formal system development with KIV. In: Maibaum, T. (ed.) FASE 2000. LNCS, vol. 1783, pp. 363–366. Springer, Heidelberg (2000). doi:10.1007/3-540-46428-X_25
5. Benatallah, B., Sheng, Q.Z., Dumas, M.: The Self-serv environment for web services composition. IEEE Internet Comput. **7**(1), 40–48 (2003)
6. Borek, M., Katkalov, K., Moebius, N., Reif, W., Schellhorn, G., Stenzel, K.: Integrating a model-driven approach and formal verification for the development of secure service applications. In: Thalheim, B., Schewe, K.-D., Prinz, A., Buchberger, B. (eds.) Correct Software in Web Applications and Web Services, Texts & Monographs in Symbolic Computation, pp. 45–81. Springer International Publishing, Cham (2015)
7. Borek, M., Moebius, N., Stenzel, K., Reif, W.: Model-driven development of secure service applications. In: 2012 35th Annual IEEE Software Engineering Workshop (SEW), pp. 62–71. IEEE (2012)
8. Borek, M., Moebius, N., Stenzel, K., Reif, W.: Model checking of security-critical applications in a model-driven approach. In: Hierons, R., Merayo, M., Bravetti, M. (eds.) Software Engineering and Formal Methods. LNCS, vol. 8137, pp. 76–90. Springer, Heidelberg (2013)
9. Borek, M., Moebius, N., Stenzel, K., Reif, W.: Security requirements formalized with OCL in a model-driven approach. In: 2013 IEEE Model-Driven Requirements Engineering Workshop (MoDRE), pp. 65–73. IEEE (2013)
10. Borek, M., Stenzel, K., Katkalov, K., Reif, W.: Abstracting security-critical applications for model checking in a model-driven approach. In: 6th IEEE International Conference on Software Engineering and Service Science (ICSESS). IEEE (2015)
11. Börger, E., Stärk, R.F.: Abstract State Machines-A Method for High-Level System Design and Analysis. Springer-Verlag, Heidelberg (2003)
12. de Castro, V., Marcos, E., Vela, B.: Representing WSDL with extended UML. Revista Columbiana de Comput. **5** (2004)
13. Dierks, T., Rescorla, E.: The Transport Layer Security (TLS) Protocol Version 1.2.IETF Network Working Group, August 2008. http://www.ietf.org/rfc/rfc5246.txt
14. Dolev, D., Yao, A.C.: On the security of public key protocols. In: Proceedings of 22th IEEE Symposium on Foundations of Computer Science, pp. 350–357. IEEE Computer Society (1981)
15. Gronmo, R., Skogan, D., Solheim, I., Oldevik, J.: Model-driven web services development. In: 2004 IEEE International Conference on e-Technology, e-Commerce and e-Service, EEE 2004, pp. 42–45. IEEE (2004)
16. Jensen, M., Feja, S.: A security modeling approach for web-service-based business processes. In: 16th Annual IEEE International Conference and Workshop on the Engineering of Computer Based Systems, ECBS 2009, pp. 340–347. IEEE (2009)
17. Katkalov, K., Moebius, N., Stenzel, K., Borek, M., Reif, W.: Modeling test cases for security protocols with SecureMDD. Comput. Netw. **58**, 99–111 (2013)
18. Mayer, P.: MDD4SOA: Model-Driven development for Service-Oriented Architectures. Ph.D. thesis, LMU (2010)
19. Menzel, M.: Model-driven security in service-oriented architectures. Ph.D. thesis. Potsdam University (2011). http://opus.kobv.de/ubp/volltexte/2012/5905/
20. Moebius, N., Stenzel, K., Reif, W.: Modeling security-critical applications with UML in the SecureMDD approach. Int. J. Adv. Softw. **1**(1), 59–79 (2008)

21. Moebius, N., Stenzel, K., Reif, W.: Generating formal specifications for security-critical applications -a model-driven approach. In: ICSE 2009 Workshop: International Workshop on Software Engineering for Secure Systems (SESS 2009). IEEE/ACM Digital Libary (2009)

22. Moebius, N., Stenzel, K., Reif, W.: Formal verification of application-specific security properties in a model-driven approach. In: Massacci, F., Wallach, D., Zannone, N. (eds.) ESSoS 2010. LNCS, vol. 5965, pp. 166–181. Springer, Heidelberg (2010). doi:10.1007/978-3-642-11747-3_13

23. Nakamura, Y., Tatsubori, M., Imamura, T., Ono, K.: Model-driven security based on a web services security architecture. In: IEEE International Conference on Services Computing, pp. 7–15. IEEE Press (2005)

24. Nolte, S.: QVT-Operational Mappings: Modellierung mit der Query Views Transformation. Springer, Heidelberg (2009)

25. Pironti, A., Pozza, D., Sisto, R.: Formally-based semi-automatic implementation of an open security protocol. J. Syst. Softw. **85**(4), 835–849 (2012). Elsevier

26. Sun Microsystems Inc.Java Card 2.2 Specification (2002). http://java.sun.com/products/javacard/

27. Thöne, S., Depke, R., Engels, G.: Process-oriented, flexible composition of web services with UML. In: Olivé, A., Yoshikawa, M., Yu, E.S.K. (eds.) ER 2002. LNCS, vol. 2784, pp. 390–401. Springer, Heidelberg (2003). doi:10.1007/978-3-540-45275-1_34

Comprehending a Service by Informative Models

Bernhard Thalheim[1]([✉]) and Ajantha Dahanayake[2]

[1] Department of Computer Science, Christian Albrechts University Kiel,
24098 Kiel, Germany
thalheim@is.informatik.uni-kiel.de
[2] Department of Computer Information Science, Prince Sultan University,
Riyadh, Kingdom of Saudi Arabia
adahanayake@pscw.psu.edu.sa

Abstract. Services are one of the main supporting facilities of modern societies. They support users in their everyday life. They provide additional features to their users. They must be useful, usable in the user environment and must correspond to the utilisation pattern of potential users. In modern applications, the user must be able to understand the service on the fly and to appreciate the utility a service provides. The user thus needs a *comprehensive* service.

Models are a mainstay of every scientific and engineering discipline. Models are typically more accessible to study than the systems. Models are instruments that are effectively functioning within a scenario. The effectiveness is based on an associated set of methods and satisfies requirements of utilisation of the model. A typical utilisation of a model is explanation, informed selection, and appropriation of an opportunity. The mental model that a user might have on a service can be supported by a specific general model of the service: the **informative model**. It generalises such conceptions as the instruction leaflet, the package or product insert, the information and direction for use, and the enclosed label.

1 The Service and Its Users

1.1 The Users and Their Comprehension Needs Before Using a Service

As IT infrastructure becomes more service-oriented and more competitive, users may select a service that best fits their needs and wishes. They need to understand what are the advantages and disadvantages of a given service compared to other similar services, what is the added value for them in a given situation, in what way a service can and should be used, what are the restrictions of the given service, which hidden features must be accepted, and what are the specific preparations before using a service, etc. Users are different and each of them has a specific understanding of the world and the given situation, i.e. also mental models of the reality.

The *mental model* is enhanced by information about the given service. The most appropriate way for information provision is a description of the service in

© Springer-Verlag GmbH Germany 2016
A. Hameurlain et al. (Eds.): TLDKS XXX, LNCS 10130, pp. 87–108, 2016.
DOI: 10.1007/978-3-662-54054-1_4

such a way that the user: may *perceive, select and organise* the best fitting service because of the user's *interest*, originating from his/her instincts, feelings, experience, intuition, common sense, values, beliefs, etc., especially the personal mental models, *simultaneously processed* by his/her cognitive and mental processes, and *seamlessly integrated* in his/her recallable mental models.
The *value* of this information lies solely in its ability to affect a behavior, decision, or outcome.

Services may be enhanced by models that represent some of the features, properties, etc. of services. The *informative model* that is introduced in the sequel is a appropriate means to support the user.

1.2   Typical Examples: Smart City and Mobile App Services

Smart cities are cities that rely on sophisticated technology to enhance the well-being of its citizens achieved through the efficient management of resources, provision of advanced infrastructure, implementation of advanced communication networks to stimulate sustainable economic growth. The fundamental aspiration is to provide a high quality of life to its citizens [18].

A *Smart City* according to [18] is a city performing in a forward-looking way in economy, people, governance, mobility, environment, and living, built on the smart combination of endowments and activities of self-decisive independent and aware citizens [18]. A smart city can be characterised on one side by [18] smart infrastructures, smart citizens, smart governance, smart mobility, smart economy, smart lifestyle, smart technology adaptation, smart service integration, and smart feedback response. Another set consisting of six characteristics is given in [10]: The first one is the utilization of networked infrastructure to improve economic and political efficiency and enable social, cultural and urban development. Second characteristics include underlying emphasis on business-led urban development. Third factor focuses on the aim to achieve the social inclusion of various urban residents in public services. The fourth characteristic stresses on the crucial role of high-tech and creative industries in relation to the established urban growth. The fifth factor has a profound focus on the role of the social and relational capital in urban development. The sixth factor then focuses on social and environmental sustainability as a major strategic component of smart cities. Characteristics five and six are the most interesting and promising ones.

In all cases, a user faces problems such as:

- *Which service can be properly used at what time, with what efforts, under which conditions, and what shall be taken for granted and what not?*
- *Which of the current tasks is supported in which way, by which data provision, and with what outcome?*
- *What kind of preparation somebody needs before accessing the service?*

Mobile application developers strive to provide user friendly, easy to use, and widely acceptable smart Mobile Apps for smart phone users. Mobile App development is a lucrative business and requires a novel approach for its design and development. Agile software development methodology is widely used with its adaptive (allowing frequent changes), incremental (revision for client at each step), cooperative (strong bond between developer and client), and straight forward (requirements are clear to modify) approach for mobile Apps development. Many Mobile Apps development approaches haven been developed, e.g. MApps [2,16,28], Mobile-D [3], MASAM [22] based on the RUP model, Hybrid [30] that supports a combination of NPD and ASD, SLeSS [11] that uses Scrum [32], or RaPiD7 [14].

Smart and mobile service target a wide auditory. Mobile services also impose a comprehension challenge to their users beside the ones we observed for smart services:

- *How the App can be intuitively used based on prior experience? For what kind of specific features?*
- *What is the utility, reliability, security, privacy, specific environment, added value, context, and provider policy of the given App?*

1.3   The Conception of a Service

Today, the service has gained recognition as the more realistic concept for dealing with complexities of cross-disciplinary systems engineering extending its validity beyond the classical information systems design and development realm [13]. In this respect the service concept combines and integrates the value created in different design contexts such as person-to-person encounters, technology enabled self-service, computational services, multi-channel, multi-device, location-based and context-aware, and smart services [36]. Therefore, the service concept reveals the intrinsic design challenge of the information required to perform a service. It emphasizes the design choices that allocate the responsibility to provide this information between the service provider and service consumer.

The service is being defined using different abstraction models with varying applications representing a multitude of definitions of the service concept [20]. The increasing interest in services requires service concept's abstraction into levels such as: business services, web services, software-as-a-service (SaaS), platform-as-a-services, and infrastructure-as-a-service [8]. Service architectures are proposed as a means to methodically structure systems [7,17,37].

A number of service notations are available in the literature, and research has looked into the service mainly from two perspectives, (a) from the low-level technological point of view and (b) from the higher abstract business point of view. These two categories of service descriptions have derived a number of service notations. Some of those main stream service notations are:

The *REA (Resource-Event-Agent) ontology* [21,26] uses as core concepts resources, economic event, and agent. The *RSS (Resource-Service-Systems) model* [29] is an adaptation of REA ontology stressing that REA is a conceptual model of economic exchange and uses a Service-Dominant Logic (SDL) [43]. The *model of the three perspectives of services* uses abstraction, restriction, and co-creation. It concentrates on the use and offering of resources [8]. The perspectives addressed by this model are: service as a means for abstraction; service as means for providing restricted access to resources; and service as a means for co-creation of value. The logics behind is the Goods Dominant Logic (GDL) model [44]. *Web service description languages* concentrate on Service-Orientated Architectures (SOAs) for web service domain. Software systems are decomposed into independent collaborating units [33]. Named services interact with one another through message exchanges. The *seven contexts of service design* [19,24,36] combine person-to-person encounters, technology-enhanced encounters, self-service, computational services, multi-channel, multi-device, and location-based and context-aware services description.

1.4   Research Issues for Service Comprehension

Services must have a description that is easy to digest by everybody. This challenge is currently not well met by most of the services. The reason is the current style of development and the non-understanding of users by developers. In reality, users have their own **mental models**. These models can be enriched in a way that the user understands the service and is able to use those services as an informed user.

We develop an approach to service specification and service modelling based on **informative models** of the service. These informative models are constructed in such a way that the user can integrate the part of the informative model into the mental model. The essential part is extracted from the informative model and incorporated into the mental model.

The *smart and mobile service challenge* is to specify informative models in such a way that users can integrate the model into their mental models. They get the right explanation for informed selection and for appropriation of the opportunity provided by the service.

1.5   The Storyline of the Paper

We start with a reconsideration of the notion of a model in Sect. 2. As a first starting point we may use an informal notion:
A model represents origins based on some abstraction with respect to the origins, and has a purpose with respect to the origins. The purpose is determined by the function property of a model: The model fulfills its function in the utilisation scenarios that are requested.

Users form their own model of a service. This formation is supported by information they request or obtain. This process is either active, proactive, or passive. Active and proactive model building should however supported be appropriate, adequate and dependable information. Therefore, we discuss modelling and comprehension of services in Sect. 3.

We propose in Subsect. 3.4 the notion of the *informative model*. This kind of a model provides all necessary information for explanation, informed selection and appropriation of a service. Whenever this model can be mapped to the mental model of users, the service becomes accepted and acknowledged.

The notion of the informative model is rather general. The mapping from the informative model to the mental models of a user can be based on a question-answer approach. We use a specification frame for such an approach. The W*H frame introduced in Sect. 4 provides a pragmatical approach to the specification of informative models.

2   The Notion of a Model

The theory of models is the body of knowledge that is concerned with the fundamental nature, function, development and utilisation of models in science and engineering, e.g. in Computer Science. In its most general sense, a model is a proxy and is used to represent some system for a well-defined purpose. Changes in the structure and behaviour of a model are easier to implement, to isolate, to understand and to communicate to others. In this section we review and refine the notion of the model that has been developed in [40–42].

2.1   Kinds of Models

Models are considered to be the third dimension of science [42]. Disciplines have developed a different understanding of the notion of a model, of the function of models in scientific research and of the purpose of the model. Models are used, for instance, as

situation models by reflecting of a given state of affairs,
perception models by reflecting of somebody's current understanding of world,
formal models that are based some formalism within a well-based formal language,
conceptual models by enhancement of formal concepts and conceptions,
experimentation models that are used as a guideline for experiments,
mathematical models that are expressed in the language of mathematics and their methods,
computational models that are based on some (semi-)algorithm,
physical models that use some physical instrument,
visualisation models that are based on a visualisation,
representation models for representation of some other notion,
diagrammatic models that are using a specific diagram language, e.g. UML, electrotechnics,
exploration models for property discovery,
heuristic models that are based on some Fussy, probability, plausibility etc. relationship, e.g. a correlation,
prediction models that are based on extrapolation assumptions such as the continuation assumptions or steadiness,
informative models that used to inform potential users about origins, and

mental models that make use of cognitive structures and operations in common use.

Models can also be distinguished on the basis of the languages they use: verbal models, tabular or relation models, hierarchical models, network or graphical models, etc. Models can be physical artifacts and virtual tools. In both cases they are *representations*.

Models can also consist of a collection of associated models, called *model suite* [38]. A model suite consists of

- a set of models with explicit *associations* among the models,
- with explicit *controllers* for maintenance of coherence of the models,
- with application schemata for their explicit *maintenance and evolution*, and
- tracers for establishment of their *coherence*.

Model suites and multi-model specification decreases the complexity of modelling due to separation of concern. Interdependencies among models must however be given in an explicit form. Models that are used to specify different views of the same problem or application must be used consistently in an integrated form. Changes within one model must be propagated to all dependent models. Each singleton model must have a well-defined semantics as well as a number of representations for display of model content. The representation and the model must be tightly coupled. Changes within any model must either be refinements of previous models or explicit revisions of such models. The change management must support rollback to earlier versions of the model suite. Model suites may have different versions.

Experimental and observational data are assembled and incorporated into models and are used for further improvement and adaptation of those models. Models are used for theory formation, concept formation, and conceptual analysis. Models are used for a variety of purposes such as:

- perception support for understanding the application domain,
- for shaping causal relations,
- for prognosis of future situations and of evolution,
- for planning,
- for retrospection of previous situations,
- for explanation and demonstration,
- for preparation of management,
- for optimisation,
- for construction,
- for hypothesis verification, and
- for control of certain environments.

Models provide a utility to research and have an added value in research, e.g. for construction of systems, for education, for the research process itself. Their added value is implicit however can be estimated based on the model capability. Models are a common culture and common practice in sciences. Each discipline has however developed its specific modelling knowledge and practice.

2.2   Representations That Are Models

A model *is a well-formed, adequate, and dependable representation of origins.* Its criteria of well-formedness, adequacy, and dependability must be commonly accepted by its community of practice within some context and correspond to the functions that a model fulfills in utilisation scenarios.

The model should be well-formed according to some well-formedness criterion. As an instrument or more specifically an artefact a model comes with its *background*, e.g. paradigms, assumptions, postulates, language, thought community, etc. The background its often given only in an implicit form. A model is used in a *context* such as discipline, a time, an infrastructure, and an application.

Models function as an instrument in some utilisation scenarios and a given usage spectrum. Their function in these scenarios is a combination of functions such as explanation, optimization-variation, validation-verification-testing, reflection-optimization, exploration, hypothetical investigation, documentation-visualisation, and description-prescription functions. The model functions effectively in some of the scenarios and less effectively in others. The function determines the *purpose* and the *objective* (or goal) of the model. Functioning of models is supported by methods. Such methods support tasks such as defining, constructing, exploring, communicating, understanding, replacing, substituting, documenting, negotiating, replacing, optimizing, validating, verifying, testing, reporting, and accounting. A model is *effective* if it can be deployed according to its objectives.

Models have several *essential properties* that qualify a representation as a model. An well-formed representation is *adequate* for a collection of origins if it is *analogous* to the origins to be represented according to some analogy criterion, it is more *focused* (e.g. simpler, truncated, more abstract or reduced) than the origins being modelled, and it sufficiently satisfies its *purpose*.

Well-formedness enables a representation to be *justified* by an *empirical corroboration* according to its objectives, by rational coherence and conformity explicitly stated through formulas, by falsifiability, and by stability and plasticity.

The representation is *sufficient* by its *quality* characterisation for internal quality, external quality and quality in use or through quality characteristics [39] such as correctness, generality, usefulness, comprehensibility, parsimony, robustness, novelty etc. Sufficiency is typically combined with some assurance evaluation (tolerance, modality, confidence, and restrictions).

A well-formed representation is called *dependable* if it is sufficient and is justified for some of the justification properties and some of the sufficiency characteristics.

2.3   Representations as Instruments in Some Utilisation Scenario

Models will be used, i.e. there is some utilisation scenario, some reason for its use, some goal and purpose for its utilisation and deployment, and finally some function that the model has to play in a given utilisation scenario. A typical utilisation scenario is problem solving. We first describe a problem, then specify the requirements for its solutions, focus on a context, describe the community of practices and more specifically the skills needed for the collaborative solution of the problem, and scope on those origins that must be considered. Next we develop a model and use this model as an instrument in the problem solving process. This instrument provides a utility for the solution of the problem. The solution developed within the model setting is then used for derivation of a solution for the given problem in the origin setting.

A similar use of models is given for models of services. Service models might be used for the development of a service system. They might be used for assessment of services, for optimisation and variation of services, for validation-verification-testing, for investigation, and for documentation-visualization. In this paper we concentrate on the *explanation, informed selection, and appropriation* use of a service model. It must provide a high level description of the service itself. This usage is typical for a process of determining whether a service is of high utility in an application. Such utilisation is based on specific utilisation pattern or more specifically on a special model that is the *utilisation model of an instrument as a model*.

2.4   Conceptional Modelling: Modelling Enhanced by Concepts

An information systems model is typically a schematic description of a system, theory, or phenomenon of an origin that accounts for known or inferred properties of the origin and may be used for further study of characteristics of the origin. *Conceptional modelling*[1] aims to create an abstract representation of the situation under investigation, or more precisely, the way users think about it. *Conceptual models* enhance models with concepts that are commonly shared within a community or at least within the community of practice in a given utilisation scenario. Concepts specify our knowledge what things are there and what properties things have. Their definition can be given in a narrative informal form, in a formal way, by reference to some other definitions, etc. We may use a large variety of semantics [34], e.g., lexical or ontological, logical, or reflective.

2.5   The Cargo of a Model

The cargo of any instrument is typically a very general instrument insert like the package insert in pharmacy or an enclosed label. It describes the instrument, the main functions, the forbidden usages, the specific values of the instrument, and the context for the utilisation model. Following [25,42], we describe the cargo by

---

[1] The words 'conceptual' and 'conceptional' are often considered to be synonyms. The word 'conceptual' is linked to concepts and conceptions. 'Conceptual' means that a thing - e.g. an instrument or representation - is characterised by concepts or conceptions. The word 'conceptional' associates a thing as being or of the nature of a notion or concept. Conceptional modelling is modelling with associations to concepts. A conceptual model incorporates concepts into the model.

a description of the *mission* of the instrument in the utilisation scenarios, the *determination* of the instrument, an *abstract declaration of the meaning* of the instrument, and a narrative explanation of the *identity* of the instrument.

The mission of a model consists of functions (and anti-functions or forbidden ones) that the model has in different utilisation scenarios, the purposes of the utilisations of the model, and a description of the potential and of the capacity of the model. The determination contains the basic ideas, features, particularities, and the utilisation model of the given instrument. The meaning contains the main semantic and pragmatic statements about the model and describes the value of the instrument according to its functions in the utilisation scenarios, and the importance within the given settings. Each instrument has its identity, i.e. the actual or obvious identity, the communicated identity, the identity accepted in the community of practice, the ideal identity as a promise, and the desired identity in the eyes of the users of the instrument.

3 Comprehending Services

3.1 The Current Way of Modelling of Smart and Mobile Services

The way of modelling specifies how the possible solution of the problem is elaborated in the form of diagrams. These models include context models (data flow diagram, sequence diagram), business models (activity diagram), business requirement models (use case diagram) and domain models (class diagram). Table 1 summarises the way of modelling followed in the mobile applications development approaches.

Table 1. The current way of modelling in mobile apps development approaches

Approach	The way Of modelling
Mobile-D [3]	Context model (Data flow diagram), Business process model (Activity diagram), Business requirement model (Use case diagram, Story cards), Domain model (Class diagram)
RaPiD7 [14]	Not defined
Hybrid methodology [30]	Context model (Data flow diagram), Business process model (Activity diagram), Business requirement model (Use case diagram, Story cards), Domain model (Class diagram)
MASAM [22]	Context model (Data flow diagram), Business process model (Activity diagram or data flow diagram), Business requirement model (Use case diagram, Story cards), Domain model (Class diagram)
SLeSS [11]	UML tool for modelling
Scrum [32]	UML tools for modelling

3.2   Mental Models

Users should properly understand, appreciate, and embody a service, i.e. in general to comprehend what is the service, how it can be used, what are the specifics of the service, what is the concept and the content of the service, what is the background of such services, what is the added value of the service, etc. Typically, users are not interested in learning the usage of such services or reading manuals. They need an easy to digest, intuitive model of the service that can seamlessly integrated into their mental models.

Mental models [27] as structures are stored in long term memory and then called upon in reasoning and those that treat them as temporary structures constructed in working memory for a specific reasoning task. Mental models are based on narratives in a broad form, e.g. visual illustrations. Narratives support a shallow simulation of events or processes as one would be in the real world. They depict abstractions based on the modeller's recognition of the situation as prototypical. It allows to draw empirical and conceptual consequences of something in the modeller's conception world. The mental model is thus an idealised, revised and polished structural and functional analog of the real world, esp. the situations perceived.

Typical examples of mental models are discourse models [23] that make explicit the structure not of sentences but of situations as users perceive or imagine them. *Discourse models* are narrative that thus consist of (1) the situation as the referent, (2) the meaning of the discourse which is often given through a description of all possible situations under current interest, and (3) an embedding into a model of the world that is judged be true. Discourse models thus embody "a representation of the spatial, temporal, and causal relationships among the events and entities of the narrative" [27].

A *situation model* is a specific mental model that is concerned with the situation observed by the modeller. It allows the modeller to generate conclusions without having to carry out extensive operations. The approximation and abstraction used in situation models allows to draw conclusions without making a deep reference to the characterisation of the situation. The conclusions are typically implicit and thus do not require deep inferential work.

3.3   Injecting Knowledge on a Service in a Mental Model of a Service

Following [15], we use the following definition:
A mental model of a service *is a relatively enduring and accessible but limited internal conceptual representation of a service whose structure maintains the perceived structure of that service.*
The *Brunswikean lens model* [9,31] uses a generalised cybernetic understanding of the interaction between a user and a system, e.g. a service. It consists of three sub-models: (1) the ends model (desired state); (2) the means model (strategies, tactics, policies, etc.); (3) the means/ends model (also called cognitive model). These three sub-models are related to the service function that is subjectively

interpreted by a human based on the interpretation abilities for those cues that are under attention.

In a simplified form, the mental model consists of a model representing the desired state (ends), the strategies and tactics to reach this state (means), and of a cognitive model a user has (means/end). The mental model is used to subjectively interpret observations relevant for the current attention. Cues are measurements for properties that are relevant and can be observed by a user in dependence on the attention a user pays. They are subjectively interpreted and used for an assessment and prediction in dependence on the means sub-model of the mental model.

Figure 1 represents this simplified understanding of an interaction between a user and a system, e.g. a service. A user needs knowledge or at least explanations about a service before the service is selected and used. The service can be used in a blind or too confiding form or can be used in an informed manner. The informative model of a service enables thus a user to select and to appropriate the service.

3.4   Informative Models

The *informative model* consists of the cargo, the description of its adequacy and dependability, and the SMART[2] and SWOT[3] statements. It informs a potential user through bringing facts to somebody's attention, provides these facts

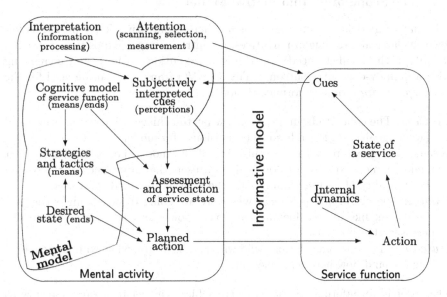

Fig. 1. The Brunswikean lens model for mental activities and their association to services

---

[2] SMART: Simple, Meaningful, Adequate, Realistic, and Trackable.
[3] SWOT: strengthes, weaknesses, opportunities, and threats.

in an appropriate form according their information demand, guides them by steering and directing, and leads them by changing the information stage and level. Based on the informative model, the user selects the origin for usage with full informed consent or refuses to use it. It is similar to an instruction leaflet provided with instruments we use. The informative model is semantically characterized by: objectivity; functional information; official information; explanation; association to something in the future; different representational media and presenters; degree of extraction from open to hidden; variety of styles such as short content description, long pertinent explanation, or long event-based description.

In the case of a service model, the informative model must state positively and in an understandable form what is the service, must describe what is the reward of a service, and must allow to reason about the rewards of the service, i.e. put the functions and purposes in a wider context (*PURE*). Informative models of a service are based on a representation that is easy-to-survey and to understand, that is given in the right formatting and form, that supports elaboration and surveying, that avoids learning efforts for their users, that provides the inner content semantics and its inner associations, that might be based on icons and pictographs, and that presents the annotation and meta-information including also ownership and usability.

We shall now explore in the sequel of what are the ingredients of such informative instruments in the case of a service model.

3.5   The Scope of an Informative Model

Informative models are often given as a coherent narrative. They may thus follow the classical storyline of narratives, i.e. introducing an exposition (context), describing the conflict (need), providing a resolution (vision), and convincing with a happily-ever-after outcome. This CoNVO frame [35] can be used for the *scope specification* of an informative model:

Context: The context clarifies the who, what-to-achieve, what-work, prioritised (larger-)goals, and the relevance dimensions of a model.

Needs: The needs refer to the utilisation of the model, i.e. to challenges that might be met, how to use, the goals of a user, the importance of the origin, the typicality of its functioning, and the satisfied demands of a user.

Vision: The vision makes it clear where we are going to, what might the user world look like after achieving the goals, how the needs are met, and the necessary background.

Outcome: The outcome clarifies who will use the origin, what will happen after using it, and how it is kept relevant.

The scope of an informative model thus enables the user to understand how to map the informative model the the mental models.

3.6   The Explanation, Selection, and Appropriation

Explanation, understanding and informed selection of a tool is one of the main usage scenarios for software models. People want to solve some problems. Services provide solutions to these problems and require a context, e.g. skills of people, an infrastructure, a specific way of work, a specific background, and a specific kind of collaboration. In order to select the right service, a model of the service is used as an *instrument for explanation and quick shallow understanding* of which service might be a good candidate, what are the strengths and weaknesses of the service under consideration, which service meets the needs, and what are the opportunities and risks while deploying such a service.

The best and simplest instrument in such usage scenario is the *instruction leaflet* or more generally as a specification of the information and directions on the basis of the *informative model*. We shall show in the sequel that this model of a service extends the cargo dimension [25] to the general notion of the informative model. Such models of a service enable people in directed, purposeful, rewarding, realistic, and trackable deployment of a service within a given usage scenario, i.e. use according to the qualities of the model [13]. After informed selection of a service, it might be used in the creation of a new work order based on the assimilation of the service into the given context, i.e. appropriation of the service.

4   Enhancing a Mental Model of Services Based on the W*H Frame

Systems and services are typically characterised by a combination of large information content with the need of different stakeholders to understand at least some system aspects. People need models to assist them in understanding the context of their own work and the requirements on it. We concentrate in this paper on the support provided by models to understand how a system works, how it can be used or should not be used, and what would be the benefit of such a model:

- to understand how a system works,
- how it can be used or should not be used, and
- what would be the benefit of such a model.

We illustrated this utilisation of models for services.

We developed a novel service model based on the W*H specification frame [13]. The W*H frame [13] provides a high-level and conceptual reflection and reflects on the variety of aspects that separates concerns such as service as a product, service as an offer, service request, service delivery, service application, service record, service log or archive and also service exception, which allows and supports a general characterization of services by their ends, their stakeholders, their application domain, their purpose and their context.

4.1   Adequacy and Dependability of Informative Models

Models are used in *explanation, informed selection, and appropriation* scenarios. The main aim of *informative models* is to inform the user according to his/her information demand and according to the profile and portfolio. The instrument steers and directs its users which are typically proactive. It supplies information that is desired or needed. Users may examine and check the content provided. Typical methods of such instruments are communication, orientation, combination, survey, and feedback methods.

Users have to be informed on what is the issue that can be solved with the instrument, what are the main ingredients of the instrument and how they are used, what is the main background behind this instrument, and why they should use this instrument. They need a quick shallow understanding of how simple, how meaningful, how adequate, how realistic, and how trackable is the instrument, i.e. how *SMART* a model is:

S*imple:* Simplicity of the model can be derived from explicit consideration for purpose. An explicit explain how the purpose would be achieved by whom, where, when, what and why. Ambiguity is not allowed.

M*eaningful:* Meaningfulness means progress of the model against the vision based on evaluation how many, how much and when it is accomplished.

A*dequateness:* Means that the model is simple, well-formed and purposeful.

R*ealistic:* It answers the questions of whether the model is doable, whether the model is worthwhile, can it be provided at the right time, match the needs and efforts of community of practice, whether it is acceptable for evolution.

T*rackable:* The model provides limits, deadlines and benchmarks. It states when, how long and when to terminate.

The user must be enabled to select the most appropriate instrument, i.e. they should know the strengths, weaknesses, opportunities, and threats of the given instrument (*SWOT*).

The SWOT and SMART evaluation is the basis for adequateness and dependability of informative models. The informative model must be analogous in structure and function to its origins. It is far simpler than the origin and thus more focussed. Its purpose is to explain the origin in such a way that a user can choose this instrument because of its properties since all demanded properties are satisfied. The selection and appropriation of an instrument by the user depends on the explanatory statement on the profile and the portfolio of the given instrument, on coherence to the typical norms and standards accepted by the community of practice, on a statement on applicability and added value of the instrument, and the relative stability of the description given. The instrument usage becomes then justified. Furthermore, the instrument must suffice the demands of such scenarios. The quality in use depends on understandability and parsimony of description, worthiness and eligibility of presented origins, and the added value it has for the given utilisation scenarios. The external quality is mainly based on its required exactness and validation. The internal quality must support these qualities. The quality evaluation and the quality safeguard are explicit statements of these qualities according to the usage scenarios, to the context, to the origins that are represented, and to community of practice.

4.2 Scenarios and Functions for Specification of SMART Models of Services

The informative model

- provides facts in appropriate form as per information demand,
- informs user, and
- guides and steers them by changing the information stages and levels.

In the case of mobile applications modelling, the informative model

- must state in an understandable way that what is the application,
- must describe the output of the application, and
- should describe the inputs of the application.

The informative model for web and mobile applications uses the W*H frame specifications [13,42]. It is is based on 23 questions as defined in [13]. Table 2 sketches the *W*H frame*. It is a novel conceptual frame for service modelling.

The W*H frame in Table 2 fulfills the conceptual definition of the service concept composing the need to serve the following purposes:

- The composition of the W*H frame consisting of *content space*, *concept space*, *annotation space*, and *add value space* as orthogonal dimensions that captures the fundamental elements for developing services.
- It reflects a number of aspects neglected in other service models, such as the handling of the service as a collection of offerings, as a proper annotation facility, as a model to describe the service concept, and as the specification of added value. It handles those requirements at the same time.
- It helps capturing and organizing the discrete functions contained in (business) applications comprised of underlying business process or workflows into interoperable, (standards-based) services.
- The model accommodates the services to be abstracted from implementations representing natural fundamental building blocks that can synchronize the functional requirements and IT implementations perspective.
- It considers by definition that the services to be combined, evolved and/or reused quickly to meet business needs.
- Finally, it represents an abstraction level independent of underlying technology.

In addition, the W*H frame in Table 2 also serves the following purposes:

- The inquiry through simple and structured questions according to the primary dimension on wherefore, whereof, wherewith, and worthiness further leading to secondary and additional questions along the concept, annotation, content, add value or surplus value space that covers usefulness, usage, and usability requirements in totality.

Table 2. The W*H specification frame for the conceptual model of a service

Service	Service name			
Concept	Ends	*Wherefore?*		
		Purpose	*Why?*	
			Where to?	
			For When?	
			For Which reason?	
Content	Supporting means	*Wherewith?*		
		Application Domain	Application are	*Wherein?*
			Application case	*Wherefrom?*
			Problem	*For What*
			Organizational unit	*Where?*
			Triggering Event	*Whence?*
			IT	*What?*
				How?
Annotation	Source	*Where of?*		
		Party	Supplier	*By whom?*
			Consumer	*To whom?*
			Producer	*Whichever?*
		Activity	Input	*What in?*
			Output	*What out?*
Added value	Surplus Value	*Worthiness?*		
		Context	Systems Context	*Where at?*
			Story Context	*Where about?*
			Coexistence Context	*Wither?*
			Time Context	*When?*

- The powerful inquiring questions are a product of the conceptual underpinning of W*H grounded within the conceptional modelling tradition in the Concept-Content-Annotation triptych extended with the Added Value dimension and further integration and extension with the inquiry system of Hermagoras of Temnos frames.
- The W*H frame is comprised of 24 questions in total that cover the complete spectrum of questions addressing the service description; (W5 + W4 + W10H +W4) and H stands for how.
- The models compactness helps to validate domain knowledge during solution modelling discussions with the stakeholders with high demanding work schedules.
- The comprehensibility of the W*H frame became the main contributor to the understanding of the domain's services and requirements.
- The model contributes as the primary input model leading to the IT-service systems projection on solution modelling.

– It contributes as the primary input model leading to the IT-service systems projection on the evaluations criteria of systems functioning on its trustworthiness, flexibility to change, and efficient manageability and maintainability.

4.3 Dimensions of Informative Models of a Service

The Content Dimension: Services as a Collection of Offerings. The service defines the what, how, and who on what basis of service innovation, design, and development, and helps mediate between customer or consumer needs and an organizations strategic intent. When extended above the generalized business and technological abstraction levels, the content of the service concept composes the need to serve the following purposes:

– Fundamental elements for developing applications;
– Organizing the discrete functions contained in (business) applications comprised of underlying business process or workflows into inter operable, (standards-based) services;
– Services abstracted from implementations representing natural fundamental building blocks that can synchronize the functional requirements and IT implementations perspective;
– Services to be combined, evolved and/or reused quickly to meet business needs; Represent an abstraction level independent of underlying technology.

The abstraction of the notion of a service system within an organizations strategic intent emphasized by those purposes given above allow us to define the content description of services as a collection of offers that are given by companies, by vendors, by people and by automatic software tools [12]. Thus the content of a service system is a collection of service offerings.

The service offering reflects the supporting means in terms of with what it means to the service's content is represented in the application domain. It corresponds to identification and specification of the problem within an application area. The problem is a specific application case that resides with an organizational unit. Those problems are subject to events that produce triggers needing attention. Those triggering events have an enormous importance for service descriptions. They couple to the solution at hand that is associated with how and what is a required IT solution.

The Annotation Dimension. According to [33], annotation with respect to arbitrary ontologies implies general purpose reasoning supported by the system. Their reasoning approaches suffer from high computational complexities. As a solution for dealing with high worst-case complexities the solution recommends a small size input data. Unfortunately, it is contradicting the impressibility of ontologies and define content as complex structured macro data. It is therefore, necessary to concentrate on the conceptualisation of content for a given context considering annotations with respect to organizations intentions, motivations, profiles and tasks, thus we need at the same time sophisticated annotation facilities far beyond ontologies. Annotation thus must link the stakeholders or parties involved and activities; the sources to the content and concept.

The Concept Dimension. Conceptional modelling aims at the creation of an abstract representation of the situation under investigation, or more precisely, the way users think about it. Conceptual models enhance models with concepts that are commonly shared within a community or at least between the stakeholders involved in the modelling process.

According to the general definition of concept as given in [41]: *Concepts* specify our knowledge what things are there and what properties things have. Concepts are used in everyday life as a communication vehicle and as a reasoning chunk. Concept definition can be given in a narrative informal form, in a formal way, by reference to some other definitions etc. We may use a large variety of semantics, e.g., lexical or ontological, logical, or reflective.

Conceptualisation aims at collection of concepts that are assumed to exist in some area of interest and the relationships that hold them together. It is thus an abstract, simplified view or description of the world that we wish to represent. Conceptualisation extends the model by a number of concepts that are the basis for an understanding of the model and for the explanation of the model to the user.

The definition of the ends or purpose of the service is represented by the concept dimension. It is the crucial part that governs the service's characterization. The purpose defines in which cases a service has a usefulness, has its usage, and has a high usability. They define the potential and the capability of the service.

The Added Value Dimension. The added value of a service to a business user or stakeholder is in the definition of surplus value during the service execution. It defines the context in which the service systems exist, the story line associated within the context, which systems must coexist under which context definitions prevailing to time. Surplus value defines the worthiness of the service in terms of time and labor that provide the Return of Investment (ROI).

4.4 Applying the W*H Frame to Informative Models of Services

The W*H conceptual model for services is applied into several real-life situations. [6] uses the W*H frame for services supporting the disease diagnosis decision support network (DDDSN) for ophthalmologists for age related macular degeneration (ARMD) treatments. This application is a crucial challenge for modern cross-disciplinary IT-service systems, as per definition they are web information systems and they are notorious for their low 'half-life' period with high velocity of evolution, migration and integration.

In [45] the W*H frame is used for modelling context-aware augmented reality mobile services. The smart mobile application developed using the W*H frame shows that it is essential to understand the needs of the human-being using it. Therefore, this W*H frame provided a description of the relationship between human needs and context information, and its role in personalization of augmented reality applications. Context-aware personalization, that is coupled with context awareness and needs prediction, predict user needs.

A third kind of applications for service modelling are scenarios for big data collection, e.g. for requirement specification of big data collection and capturing services [1], for collecting data according to scenarios of interest for analysis of (real-time) decision support, and for the reduction of unnecessary or garbage data collection Garbage is a huge problem for big data in terms of storage, transportation, and analytic time for (real-time) decision support. and for requirements acquisition tool architecture design for social media big data collection [5] for the data collection to accelerate the analytics tasks.

Another smart mobile application for service modelling is health care management mobile services [4]. In this particular situation the Diabetic Medicare service is modelled using the W*H frame. The App assists the diabetic patients in managing diabetes by providing them timely information and advice regarding the medication, the diet etc.

Further, the W*H frame was taken into consideration in the domain of methodology engineering for modelling smart modelling services for mobile application development approaches [4]. It puts forward the use of the Simple, Meaningful, Adequate, Realistic, and Trackable (SMART) models. Thereby, developers have at their disposal a powerful model to create the mobile application services in a more systematic and structured manner.

5 Conclusion

There are many other usage models for services. This paper elaborated the *explanation, informed selection, and appropriation* usage model for a service. Other usage models of an instrument as a model are, for instance, optimization-variation, validation-verification-testing, understanding, extension and evolution, reflection-optimization, exploration, documentation-visualization, integration, hypothetical investigation, and description-prescription usage models. We introduced in this paper a general notion of the model and showed what makes description or specification a service to be become a model of the service.

References

1. Al-Najran, N., Dahanayake, A.: A requirements specification framework for big data collection and capture. In: Morzy, T., Valduriez, P., Bellatreche, L. (eds.) ADBIS 2015. CCIS, vol. 539, pp. 12–19. Springer, Heidelberg (2015). doi:10.1007/978-3-319-23201-0_2
2. Abrahamsson, P., Warsta, J., Siponen, M.T., Ronkainen, J.: New directions on agile methods: a comparative analysis. In: Proceedings of the 25th International Conference Software Engineering 2003, vol. 6, pp. 244–254 (2003)
3. Abrahamsson, P., Hanhineva, A., Hulkko, H., Ihme, T., Jäälinoja, J., Korkala, M., Koskela, J., Kyllönen, P., Salo, O.: Mobile-D an agile approach for mobile application development. Int. J. Serv. Ind. Manag., 174–175 (2004)
4. Alsabi, E.: Methodology Engineering for Mobile Application Development. Master of Software Engineering thesis. Prince Sultan University, KSA (2016)

5. AlSwimli, M., Dahanayake, A.: Tool for decision Backed Big Data Capture (2015, Submitted to Information Systems Frontiers)
6. Amarakoon, S., Dahanayake, A., Thalheim, B.: A framework for modelling medical diagnosis and decision support services. Int. J. Digit. Inf. Wirel. Commun. (IJDIWC) **2**(4), 7–26 (2012)
7. Arsanjani, A., Ghosh, S., Allam, A., Abdollah, T., Ganapathy, S., Holley, K.: SOMA: a method for developing service-oriented solutions. IBM Syst. J. **47**(3), 377–396 (2008)
8. Bergholtz, M., Andersson, B., Johannesson, P.: Abstraction, restriction, and co-creation: three perspectives on services. In: Trujillo, J., et al. (eds.) ER 2010. LNCS, vol. 6413, pp. 107–116. Springer, Heidelberg (2010). doi:10.1007/978-3-642-16385-2_14
9. Brunswik, E.: Perception and the Representative Design of Psychlogical Experiments. University of California Press, Berkeley (1956)
10. Caragliu, A., Del Bo, C., Nijkamp, P.: Smart cities in Europe. J. Urban Technol. **18**(2), 65–82 (2011)
11. Da Cunha, T.F.V., Dantas, V.L.L., Andrade, R.M.C.: SLeSS: a scrum and lean six sigma integration approach for the development of software customization for mobile phones. in: Proceedings of 25th Brazilian Symposium on Software Engineering, SBES 2011, pp. 283–292 (2011)
12. Dahanayake, A.: CAME: An Environment for Flexible Information Modeling. Ph.D. Dissertation. Delft University of Technology, the Netherlands (1997)
13. Dahanayake, A., Thalheim, B.: Co-design of web information systems. In: Correct Software in Web Applications and Web Services, pp. 145–176. Springer, Wien (2015)
14. Dooms, K., Kylmäkoski, R.: Comprehensive documentation made agile – experiments with RaPiD7 in philips. In: Bomarius, F., Komi-Sirviö, S. (eds.) PROFES 2005. LNCS, vol. 3547, pp. 224–233. Springer, Heidelberg (2005). doi:10.1007/11497455_19
15. Doyle, J.K., Ford, D.N.: Mental models concepts for system dynamics research. Syst. Dyn. Rev. **14**, 3–29 (1998)
16. Education, I.J.M., Science, C., Khalid, A., Zahra, S., Khan, M.F.: Suitability and contribution of agile methods in mobile software development, pp. 56–62, February 2014
17. Erl, T.: SOA: Principles of Service Design. Prentice-Hall, Englewood Cliffs (2007)
18. Giffinger, R., Fertner, C., Kramar, H., Kalasek, R., Pichler-Milanovic, N., Meijers, E.: Smart cities ranking of European medium-sized cities. Vienna University of Technology (2007)
19. Glushko, R.J.: Seven contexts for service system design. In: Maglio, P.P. et al. (eds.) Handbook of Service Science, Service Science: Research and Innovations in the Service Economy. Springer Science+Business Media LLC, New York (2010). doi:10.1007/978-1-4419-1628-0_11
20. Goldstein, S.M., Johnston, R., Duffy, J.-A., Rao, J.: The service concept: the missing link in service design research? J. Oper. Manage. **20**, 121–134 (2002). Elsevier
21. Hurby, P.: Model-Driven Design of Software Applications with Business Patterns. Springer, Heidelberg (2006)
22. Jeong, Y.J., Lee, J.H., Shin, G.S.: Development process of mobile application SW based on agile methodology. Int. Conf. Adv. Commun. Technol. ICACT **1**, 362–366 (2008)
23. Johnson-Laird, P.N.: Mental models. In: Posner, M. (ed.) Foundations of Cognitive Science, pp. 469–500. MIT Press, Cambridge (1989)

24. Maglio, P., Srinivasan, S., Kreulen, J., Spohrer, J.: Service systems, service scientists, SSME, and innovation. Commun. ACM **49**(7), 81–85 (2006)
25. Mahr, B.: Zum Verhältnis von Bild und Modell. In: Visuelle Modelle, pp. 17–40. Wilhelm Fink Verlag, München (2008)
26. McCarthy, W.E.: The REA accounting model: a generalized framework for accounting systems in a shared data environment. Account. Rev. **57**, 554–578 (1982)
27. Nersessian, N.J.: Thought experimenting as mental modeling. In: Proceedings of the Biennial Meeting of the Philosophy of Science Association, vol. 2, pp. 291–301. The University of Chicago Press (1992)
28. Nosseir, A., Flood, D., Harrison, R., Ibrahim, O.: Mobile development process spiral. In: 2012 Seventh International Conference on Computer Science, Engineering, pp. 281–286 (2012)
29. Poels, G.: The resource-service-system model for service science. In: Trujillo, J., et al. (eds.) ER 2010. LNCS, vol. 6413, pp. 117–126. Springer, Heidelberg (2010). doi:10.1007/978-3-642-16385-2_15
30. Rahimian, V., Ramsin, R.: Designing an agile methodology for mobile software development: a hybrid method engineering approach. In: Second International Conference on Research Challenges in Information Science (2008)
31. Richardson, G.P., Andersen, D.F., Maxwell, T.A., Steward, T.R.: Foundations of mental model research. In: Proceedings of the International System Dynamics Conference, pp. 181–192 (1994)
32. Scharff, C., Verma, R.: Scrum to support mobile application development projects in a just-in-time learning context. In: Proceedings of 2010 ICSE Workshop Cooperative and Human Aspects of Software Engineering, CHASE 2010, pp. 25–31 (2010)
33. Schewe, K.-D., Thalheim, B.: Development of collaboration frameworks for web information systems. In: IJCAI 2007 (20th International Joint Conference on Artificial Intelligence, Section EMC 2007 (Evolutionary models of collaboration)), Hyderabad, pp. 27–32 (2007)
34. Schewe, K.-D., Thalheim, B. (eds.): Semantics in Data and Knowledge Bases, vol. 4925. Springer, Heidelberg (2008)
35. Shron, M.: Thinking with Data. O'Reilly, Sebastopol (2014)
36. Spohrer, J., Maglio, P.P., Bailey, J., Gruhl, D.: Steps towards a science of service systems. IEEE Comput. **40**, 71–77 (2007)
37. Stojanovic, Z., Dahanayake, A.: Service - Oriented Software Systems Engineering: Challenges and Practices. Idea Group Publishing, USA (2004)
38. Thalheim, B.: Model suites for multi-layered database modelling. In: Information Modelling and Knowledge Bases XXI. Frontiers in Artificial Intelligence and Applications, vol. 206, pp. 116–134. IOS Press (2010)
39. Thalheim, B.: Towards a theory of conceptual modelling. J. Univ. Comput. Sci. **16**(20), 3102–3137 (2010). http://www.jucs.org/jucs_16_20/towards_a_theory_of
40. Thalheim, B.: The conceptual model ≡ an adequate and dependable artifact enhanced by concepts. Information Modelling and Knowledge Bases. XXV of Frontiers in Artificial Intelligence and Applications, vol. 260, 241–254. IOS Press (2014)
41. Thalheim, B.: Models, to model, and modelling - towards a theory of conceptual models and modelling - towards a notion of the model. Collection of recent papers (2014). http://www.is.informatik.uni-kiel.de/~thalheim/indexkollektionen.htm
42. Thalheim, B., Nissen, I. (eds.): Wissenschaft und Kunst der Modellierung. De Gruyter, Ontos Verlag, Berlin (2015)

43. Vargo, S.L., Lusch, R.F.: Evolving to a new dominant logic for marketing. J. Mark. **68**, 1–17 (2004)
44. Vargo, S.L., Maglio, P.P., Akaka, M.A.: On value and value co-creation: a service systems and service logic perspective. Eur. Manag. J. **26**, 145–152 (2008)
45. Yahya, M.: A Context-Aware Personalization Model for Augmented Reality Applications. Master of Software Engineering Thesis. Prince Sultan University, KSA (2016)

Providing Ontology-Based Privacy-Aware Data Access Through Web Services and Service Composition

Sven Hartmann[1]([⊠]), Hui Ma[2], and Panrawee Vechsamutvaree[2]

[1] Clausthal University of Technology, Clausthal-Zellerfeld, Germany
sven.hartmann@tu-clausthal.de
[2] Victoria University of Wellington, Wellington, New Zealand
hui.ma@ecs.vuw.ac.nz, panraweev@gmail.com

Abstract. Web services have emerged as an open standard-based means for publishing and sharing data through the Internet. Whenever web services disclose sensitive data to service consumers, data privacy becomes a fundamental concern for service providers. In many applications, sensitive data may only be disclosed to particular users for specific purposes. That is, access to sensitive data is often restricted, and web services must be aware of these restrictions such that the required privacy of sensitive data can be guaranteed. Privacy preservation is a major challenge that has attracted much attention by researchers and practitioners. Hippocratic databases have recently emerged to protect privacy in relational database systems where the access decisions, allowed or denied, are based on privacy policies and authorization tables. In particular, the specific purpose of a data access has been considered. Ontologies has been used to represent classification hierarchies, which can be efficiently accessed via ontology query languages. In this paper, we propose an ontology-based data access model so that different levels of data access can be provided to web service users with different roles for different purposes. For this, we utilize ontologies to represent purpose hierarchies and data generalization hierarchies. For more complex service requests that require composite web services we discuss the privacy-aware composition of web services. To demonstrate the usefulness of our access control model we have implemented prototypes of financial web services, and used them to evaluate the performance of the proposed approach.

1 Introduction

Service-oriented computing promises rapid development of software systems by composing many distributed inter-operating autonomous services on the web. Web-based software systems in areas like e-health, e-government, e-science or e-commerce frequently require the exchange of sensitive data among invoked web services. Examples include electronic health records, contact details, salary statements, payment details, location data, and behavioral data (that allows one to track where customers have been and what they were doing).

© Springer-Verlag GmbH Germany 2016
A. Hameurlain et al. (Eds.): TLDKS XXX, LNCS 10130, pp. 109–131, 2016.
DOI: 10.1007/978-3-662-54054-1_5

Privacy preservation (in compliance to legal regulations) is a major economic concern for many organizations and enterprises using such software. Software systems that are used for processing sensitive data of customers are expected to be privacy-aware, that is, to provide suitable mechanisms for protecting sensitive data from disclosure and misuses [1].

The preservation of data privacy for shared data has attracted considerable research interest. View based approaches are proposed to define fine-grained access control models. However, there are several limitations of view-based approaches. Firstly, it is not scalable, which means that administrators have to create views for each user. Secondly, access of data is binary, i.e., either allowed or denied, which is too restricted for some queries that only requires generalized information of the required data. Thirdly, views need to be changed whenever policies are changed [2]. Hippocratic databases have been proposed in [3] to define and enforce privacy rules in database systems by adding purposes to tables. The use of Hippocratic databases for limiting data disclosure has been studied in [4,5]. Note that the access of data in a Hippocratic database is binary, i.e. either allowed or denied. Providing only binary access of data is not flexible. In [6], the Hippocratic database mechanism [3] is extended by supporting hierarchical purposes as well as distributed and minimal sets of authorizations. In [4], Hippocratic databases are enhanced to enforce minimal disclosure of data so that data owners can control who is allowed to see their personal information and under what circumstances. [7] extends the work in [3] by providing different access levels using generalization hierarchies, of which the data elements are generalized and stored in the metadata table in the database. Generalization hierarchies are also used in other work for privacy awareness [8–10]. K-anonymity, which is achieved by generalizing or suppressing values in the specific tuples of the table, is a well-known privacy protection using the generalization technique [11–13]. However, k-anonymity generalization often leads to an information loss [14,15], i.e., the reduction of the utility in the masked data [16]. To minimize information loss, [15] proposes to use ontology concepts for data anonymity and classification hierarchy. In [17], it is proposed to use ontologies for ensuring authorized access of user profiles among third parties. However, it is not discussed how authorized accesses can be ensured.

In [18,19], ontology-based implementations of the role-based access control model (RBAC) and the attribute-based access control model (ABAC) are presented. To achieve a fine-grained access control and maximize the usability of customer data, [20] proposed a role-involved purposed-based access control model, where a conditional purpose is defined as the intention of data accesses or usage under certain conditions. The access model is based on purposes that are defined by purpose hierarchies. Access decisions are determined by implied purposes that can be computed from the purpose hierarchy. However, there is no performance evaluation of the proposed access model. The computation of the implied purposes can be expensive and therefore affect the overall performance of the data access control.

The privacy of data accessed through web services has been addressed in the research literature. [1] proposes to use Hippocratic databases to provide data for web services so that related information will be released only to authorized users and for a limited period, based on purpose. In [21], the authors present semantic based privacy framework for web services. However, there are only three permission level rules that are imposed on a data element for a given service class, Free, Limited and NotGiven. Also, generalization of the data elements was not considered.

Ontologies are explicit formal specifications of the terms in the domain and relations among them [22]. They have been widely used as concept hierarchies [14,23]. Ontologies can be queried with special query languages, e.g. SPARQL. Various tools (e.g., Protégé) are available to develop ontologies. To avoid expensive computation of implied purposes as in [20] and to minimize information loss we will employ ontologies in this paper for constructing purpose hierarchies and data generalization hierarchies.

In this paper we propose an ontology-based access model that provides different levels of data access based on roles of service users and the purposes of their requests to access data. For this we will employ ontologies to capture privacy rules and generalization hierarchies of data. We will evaluate the performance of our access model with experiments and report on prototypes of financial web services that have been implemented as a proof-of-concept.

Organization. This paper is organized as follows. Section 2 introduces a motivating example that will be used throughout the paper to illustrate our objectives. Section 3 provides details on privacy requirement specification and access modeling that incorporates different access hierarchies. Upon this basis, we present our role-purpose ontologies and data generalization ontologies in Sect. 4. In Sect. 5 we discuss the implementation of our proposed ontology-based access control model. In Sect. 6 we explore the privacy-aware composition of web services in the context of our access control model. Section 7 is devoted to the evaluation of our proposed access control model. Finally, Sect. 8 concludes the paper and proposes open questions for future research.

2   A Motivating Example

Assume we have a web service that provides personal data of customers (e.g., contact details, health status, or financial information) to users with different roles for different purposes. The customer as the owner of the data likes to share their personal data only with users who assume a particular role and want to access it for a specific purpose. That is, for the customer's personal data different usage rules apply for different roles and different purposes. The usage rules are declared in the privacy policy. Table 1 outlines a customer's microdata stored in the Customer table.

Assume the privacy policy of using Customer data includes the following usage rules:

Table 1. Customer

Firstname	Lastname	DoB	Gender	Salary	Deposit	Illness	Postcode
Matt	Rovel	1960-08-18	Male	$2400	$8520	Pericarditis	6140

1. For a user assuming the role of system administrator with the purpose of maintaining customers personal data in the system, the customer allows her to view the customer's entire data.
2. For a user assuming the role of insurance officer with the purpose of providing insurance advices, the customer likes to provide information about his salary range, the birth year and the amount of deposit in the bank. That is, the customer opts to provide generalized information that is still sufficient for the user to accomplish the specified purpose. For that, instead of providing the exact value of his salary or deposit, respectively, these values are generalized to an interval, and only the respective interval is released to the user.
3. For a user assuming the role of bank agent with the purpose of providing credit card service, the customer agrees to provide approximate information of his deposit amount in the bank.
4. For a user assuming the role of bank agent with the purpose of providing home loan service, the customer is willing to provide approximate information about his salary.

It is widely accepted that different users have different needs to access sensitive data (e.g., personal data of customers), often depending on the particular role the user assumes. For example, an insurance officer might need more detailed information about the health status of a customer than a bank agent does. While the customer might opt to give the insurance officer access to his entire health record, he might not be willing to grant the same level of detail to the bank agent. Rather, some generalized version of the information might suffice for the bank agent.

However, it is not only the role that matters. For the same role different usage rules for accessing sensitive data may apply depending on the specific purpose for which the user wants to access the data. For example, the customer might allow the insurance officer to access his health record in full detail for the purpose of purchasing a new health insurance, while the customer grants the insurance office less detailed access for the purpose of claiming a discount on premiums, or even no access for the purpose of updating contact details. Considering the variety of requirements for restricting access to sensitive data through web services, we can see the demand for an access model of web services that grants data access depending on the particular role of the user who is requesting the service, and on the specific purpose of accessing the data. Such an access model is the basis for effectively preserving the privacy of data by web services.

3   Privacy Model

As we seen from the example in the above section, data access level for users with different roles and purposes should be different so that only the information needed for the roles and purposes are provided to the users. In the following we will first specify privacy requirement from data providers or data owners and then we present an access model that be used to realized privacy requirement. Our model defines data access level based on the roles and purposes of service consumers.

3.1   Privacy Requirement Specification

As we see from the example in the above section, data accesses for different roles and different purposes should be different so that only the information needed are provided to service users. It is important to understand and record privacy requirements of data providers. For this we need privacy requirement specification. We can define privacy requirement specification with a form of $P((r,p), A\{a_1 : l_1, \ldots, a_n : l_n\})$, where r is a role of the user, p refers to the purpose of data, $a_i : l_i$ refers to an attribute and its access level, i.e., write, read and the granularity of data to be read.

For example, the expression $P((Bank, CreditCard), A\{Deposit : level\_3, Age : level\_5\})$, meaning that a bank agent can access customer $deposit$ at $level\_3$ and age at $level\_5$. Data generalization hierarchies can be used to define the granularity of data returned to a bank agent.

Once a privacy requirement is specified by the owner of data, it can then be assigned to individual users. To do this, predicates can be defined to assign a user to a role and some purposes. We can use an expression $\lambda(u, (r, p))$. The relationship of role and purpose can be defined with role-purpose hierarchy.

For example, we can assign a user $Paul$ to role $Bank$ who can use web services with purposes $CreditCard$ and $HomeLoan$, $\lambda(Paul, (Bank, CreditCard))$, and $\lambda(Paul, (Bank, Homeloan))$.

3.2   Access Modelling

Once privacy requirement has been specified, we need to decide how to set up access of data so that privacy requirements can be satisfied. For that we propose to use some access hierarchy using ontologies [24]. As we see from practice users have different levels of data accesses, depending on their roles and the tasks they need to perform.

To allow a suitable service to be provided to a service requester, data access level of services should provide an adequate description, which requires the definition of an ontology [24]. That is, we need a *terminological knowledge* layer (aka TBox in description logics) describing purposes and roles (or relationships) among them. This usually includes a subsumption hierarchy among concepts, i.e. roles and purposes.

For example, consider a set of roles accessing data via services, e.g., {Admin, Bank, Insurance}. Assume the access level of bank agents is more restricted than administrators. If a user is assigned a role of administrator, then the user can access all the data that a bank agent can access. In the mean time, the purpose of a user to access a service determines the granularity of data presented to the user. To capture the semantic relationship among different role-purpose pair, we introduce the notion of role-purpose subsumption, noted $\sqsubseteq_{rp}$. For instance, the following subsumption can be specified: $Bank\_HomeLoan \sqsubseteq_{rp} Bank$. Note that role-purpose subsumption is different from the typical subsumption of domain concepts, noted by $\sqsubseteq$.

To define data granularity we can use data generalization hierarchy, similar as we define role-purpose hierarchy. For each data item, different granularity levels of data can be defined. Also, we know that access operations are related to each other. For example, if we are allowed to *write* an attribute we are allowed to *read* and *update* the attribute. In this paper we treat *read* as the same as $Level_1$, meaning the original data can be presented. Therefore, we need to model relationship between different data access operations.

In the process of service composition with privacy being preserved the following conditions need to be satisfied:

1. For a given pair of (role, purpose), for an attribute a if the access level is higher that $Level_1$. then a generalized data need to be retrieved from data generalization hierarchy and presented with the service.
2. For a given pair of (role, purpose), if there is no data generalization level on any attribute, data should be returned with a level defined for its parent on the role-purpose hierarchy.

4   Role, Purpose, and Generalization

The privacy policies are declared in terms of usage rules that define which data can be used by whom for which purpose and for how long. In this paper, ontologies are used to manage the privacy policies in term of usage rules and generalization. The usage rules are assigned to each attribute in the database tables. By this, the access of each attribute depends on the user's roles and the purpose of the data access.

Purposes of data access often have hierarchical associations among them [20]. In this paper we use ontologies to construct purpose hierarchies. Ontology defines concepts used to describe and represent a knowledge area. In particular, ontology can be used to construct concept hierarchies. Ontologies are constructed by using formal languages, which allow to include reasoning rules that support the processing of that knowledge.

In this section we will present an ontology-based access model for web services. As mentioned in Sect. 1, by using ontologies to capture data access purposes hierarchies we do not need to define the set of all the implied purposes as in [20]. Also, we will use ontologies to define data generalization hierarchies, which can be used to define different levels of data access. The remaining of the section starts with defining role-purpose ontology hierarchies, followed by defining data generalization ontology hierarchies.

4.1 Defining Service Access Roles and Purposes

Roles and purposes are important factors when deciding data shared through web services. Access decisions are based on roles and purposes. Our access model makes use of an ontology hierarchy to capture roles and purposes of service users together with relationships between roles and purposes. The root of the hierarchy is *role*. Each *role* may have different access purposes. Therefore, we define its access purposes as a subconcept *role_purpose* of *role*.

Consider our example in Sect. 2. We can define *Admin_MaintCustomer* as subconcept of *Admin*, and *Bank_HomeLoan* as a subconcept of *Bank*. Figure 1 shows an example of a role-purpose ontology hierarchy.

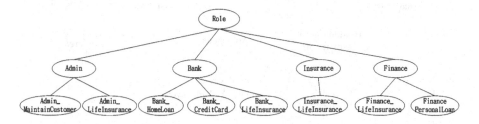

Fig. 1. Roles and purposes hierarchy

Using the concept of *role_purpose* hierarchy we can define access rules for each attribute in a database. Therefore usage rules can be represented by a triple: ⟨*role_purpose, attribute, access_level*⟩.

4.2 Defining Data Generalization Hierarchies

Generalization can help the organizations provide better services to customers without unnecessary violation to individual privacy. Generalization is a process of replacing a value with a less specific but semantically consistent value [11].

To provide more flexible access to data, generalization has been used to provide different levels of data access without actual data (microdata). In this section we use ontology to define generalization hierarchies for data. Using generalization can avoid unnecessary revealing of sensitive data, e.g., customers' salary, date-of-birth, illness and address [25]. Using ontology has been proved by [26] that it makes the generalization more meaningful.

There are various approaches of generalization in the literature. How to build generalization is not in the scope of this research. In this paper we apply the generalization approach proposed in [27] due to its simplification. Information of attributes of a table is normally divided into two types, *categorical informa-tion* and *numeric information*. For the categorical information, such as diseases, generalization for these data elements is typically described by a taxonomy. The leaf nodes ($k = 1$) depict all the possible values of data elements. Generalization for numeric data, such as ages, salary and deposit, is done by discretization of its values into a set of disjoint intervals.

For example in Fig. 2, the leaf nodes of disease hierarchy are *endocarditis*, *pericarditis*, *myocarditis* and *cardiomyopathy*. These diseases can be grouped into *heart functional malfunctions*, which is the parent level of the leaf nodes. Further, *heart functional malfunction* can be generalized to *heart disease*. Figure 3 shows the example of generalization hierarchy of numeric data, deposit and salary.

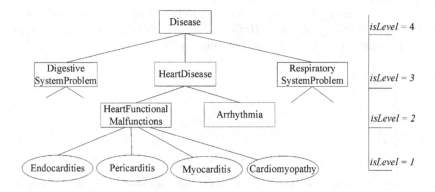

Fig. 2. Example data generalization hierarchy 1

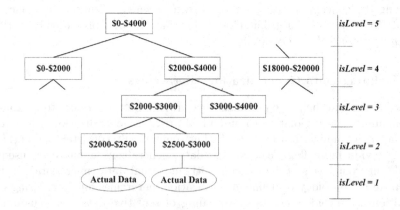

Fig. 3. Example data generalization hierarchy 2

In the ontology, the generalization level k is defined by using the *isLevel* datatype property. The microdata is defined to be at *isLevel* = 1 and is stored in the database. The ontology stores the data of generalization level $k > 1$. The more generalized the data is, the higher the value of *isLevel* is. Different (*role*, *purpose*) have different data access levels of generalization hierarchies.

4.3   Defining Access Levels

As we see above, data access rules can be defined by assigning an access level to each (*role-purpose*, *attribute*) pair. Accessing data is not only via SELECT operation but can also via other operations, i.e. INSERT, UPDATE and DELETE. Therefore access models should support multiple DML operations. In our access model each attribute of a table is assigned an access level. We will use three access levels in our access model, *canReadBy*, *isLevel* and *canWriteBy*.

- *canReadBy* refers to SELECT operation. If the attribute is assigned *canReadBy* for a *role*, it means that the *role* can access the value of the attribute with all *purpose* values that are subtypes of the *role*. If the attribute is assigned *canReadBy* for a (*role*, *purpose*) pair value, it means that the *role* can access the attribute value only for the *purpose*.
- *isLevel_X*, where $X \in \{1, 2, \ldots, n\}$, is the level of generalization. The attributes are assigned *isLevel_X* property to indicate the level of generation. The higher value X is, the higher level the generalization is.
- *canWriteBy* refers to INSERT, UPDATE and DELETE operations. We group these operations together since if a role can insert a value, it should also be able to update and delete the value of the attribute.

Note that *isLevel_1* has an equivalent property as *canReadBy*. Also, an attribute that has the access type *canWriteBy* should also have the access type *canReadBy*. Therefore we can model *canReadyBy* as a subtype of *canWriteBy*.

Often, the accesses of data need to be with a certain period of time. Limited retention time is one of the principles of Hippocratic database systems. It ensures that data is retained only as long as necessary for the fulfillment of a given service request. In our approach we use *hasRetention* property to set up retention time.

4.4   An Example of Access Rules Based on Roles and Purposes

For example, we can define attribute access levels for all attributes in table Customer in Sect. 2 as shown in Table 2. It shows attribute access levels for the *LifeInsurance* purpose for different roles.

To ensure usage rules on attribute *illness* we assign role *Insurance*, purpose *LifeInsurance* and access level *isLevel_2*. It means that insurance officers can view generalized data of customer personal_illness. Similarly, on the same attribute we assign role-purpose *Insurance_LifeInsurance* and role-purpose *Bank_LifeInsurance* we assign access level *isLevel_3*. Ontologies can be implemented with RDF files and queried by SPARQL.

Ontologies are used to define the terms used to describe and represent an area of knowledge. Ontologies are constructed by using formal languages. Languages

Table 2. Assigning access levels to attributes for *LifeInsurance* purpose

role	Firstname	Lastname	DoB	Gender	Salary	Deposit	Illness
Admin	*canReadBy*	*canReadBy*	*isLevel_1*	*canReadBy*	*isLevel_1*	*isLevel_1*	*isLevel_1*
Insurance	*canReadBy*	*canReadBy*	*isLevel_4*	*canReadBy*	*isLevel_3*	*isLevel_3*	*isLevel_2*
Bank	*canReadBy*	*canReadBy*	*isLevel_4*	*canReadBy*	*isLevel_4*	*isLevel_4*	*isLevel_3*

can allow to include reasoning rules that support the processing of that knowledge. Ontology languages are based on a well defined semantics which enables to map concepts from various perspectives. One of the advantages of ontologies is the support of concept hierarchies. In this paper, we use this advantage to implement ontological generalization hierarchies.

In this paper, ontologies are used to manage the privacy policies in term of usage rules and generalization. The usage rules are assigned to each attribute in the database tables. By this, each attribute has limited accesses, dependent on the users' role and purpose of the access.

For different purpose our web-based system will provide different data via different web services. Web services are used not only for retrieving information (SELECT operation), but also for inserting, updating and deleting data in the database. However, different roles have different access type. For example, administrators can select, insert and update customers' health status while insurance agents can only select, and bank agents cannot even view this kind of information.

Access decision is based on roles and purposes. In our ontology model, roles and purposes are stored together in the format of *role_purpose* such as *Admin_CreditCard*, *Bank_HomeLoan* and *Insurance_LifeInsurance*. Figure 4 shows the hierarchy of how we design roles and purposes in the ontology. From Fig. 4, we can see that only accessible purposes are listed as a subclass of the roles. For example, insurance role can access only LifeInsurance web service.

5   Ontology-Based Web Services Access Model

In this section we first present an architecture of a web-based system that provide data access via web services to service users. We will then provide a detailed description of an ontology-based access model that is used by the architecture.

5.1   Data Access via Web Services: An Architecture

Figure 5 shows the components and the relationships among the components in the architecture. There are six components in the architecture: web client, servlet, Credential Checking Web Service (CCWS), Data Gathering Web Service (DGWS), database and ontology.

The web client provides interfaces for users to access data via web services. CCWS is used to check user's authority. Once users are logged in, the system

Fig. 4. Roles and purposes hierarchy

checks their roles in the Ontology. If a user is authorized, the system returns user's role and a list of accessible web services. DGWS checks which tables and attributes can be accessed by a role with a purpose at which level. If accessible attributes need to be generalized before they are displayed to the user, it retrieve generalized values of the attributes that the $(role, purpose)$ can access. Ontology stores access rules while the database stores microdata of customers. Using this architecture we can provide different levels of data access via web services based on roles and purposes. CCWS, DGWS and ontology are used for access control of data in the database and are therefore the core of this architecture.

5.2   A Web Service Access Model

The access model used in the architecture above is shown in Fig. 6. The access model can be described with the following steps:

1. a user logs in with $username, password$, which are passed to CCWS to check user's credential by using a SPARQL query (Query 1) to query the role-purpose ontology. If the user has access right, the query will return user's role $(role)$ and accessible purposes acc_pur.
2. The user selects a customer name cus_name, and a purpose acc_pur for accomplishing his/her task for the customer. Using values of $role$ and acc_pur DGWS sends a SPARQL query (Query 2) to retrieve accessible attributes acc_attr from the ontology.
3. Take input acc_attr, DGWS creates a SQL query to retrieve the customer's microdata $microdata$ from a table $\$tables$ in the database.
4. Take attributes list acc_attr as input, DGWS retrieves a set of attributes gen_attr that needs to be generalized before displayed to the user. If such

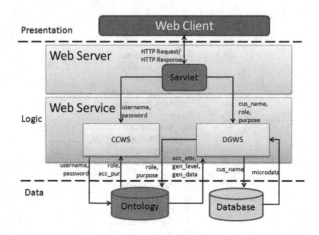

Fig. 5. Architecture of the prototype

attributes *gen_attr* exist, DGWS uses a SPARQL (Query 4) to get the generalization level *gen_level* from the ontology.

5. Take input *gen_level*, DGWS sends a SPARQL (Query 5) to get the generalization value *gen_data* for attribute *gen_attr*.

6. The results are displayed in a web service for the user.

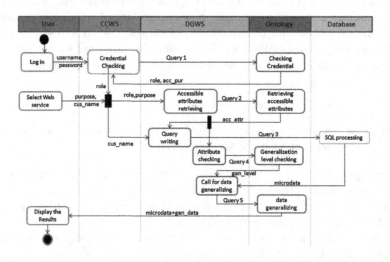

Fig. 6. An ontology-based web service access model

Lets $x and ?y be parameters used in a query, where x is a value that we pass to the web services and y is a parameter in the query. In the following we show the SPARQL queries used in the steps above.

Query 1. (login) SPARQL query for credential checking.
PREFIX Fi: <http://www.owl-ontologies.com/Finance2#>
PREFIX db: <http://biostorm.stanford.edu/db_table_classes/
          DSN_jdbc.mysql.//localhost.3306/db#>
PREFIX rdfs: <http://www.w3.org/2000/01/rdf-schema#>
SELECT ?user ?role ?acc_pur
WHERE { ?user db:user_account.user_id $username$.
          ?user db:user_account.password $password$.
          ?user Fi:hasRole ?role.
          ?acc_pur rdfs:subClassOf ?role. }

Once a user has logged in, the system shows only the services that can be accessed by the particular role. Each of the services is for one purposes. The user can select a web service for accomplishing his/her purpose. Both role and purpose values are used by Query Writing Module to retrieve accessible attributes (*accessible_attr*) from ontology. Query 2 is used to retrieve the accessible attributes.

Query 2. (getAttr) SPARQL query for retrieving accessible attributes.
PREFIX Fi: <http://www.owl-ontologies.com/Finance2#>
SELECT ?accessible_attr
WHERE { ?accessible_attr Fi:canReadBy $role_purpose$. }

After we have got the list of accessible attributes, Query Writing module creates an SQL query (Query 3) to retrieve the microdata (*microdata*) from the database. The $tables$ parameter in Query 3. is retrieved at the same time as we get the attributes list. The attributes list is also checked by Data Generalizing module whether there is an attribute that needs to be generalized before displayed to the user. We use *gen_attr* parameter to refer to the attributes that needed to be generalized. If such attributes exist, the Data Generalization will get a generalization level (*gen_level*) from ontology. The generalization level is depended on three values, *role*, *purpose* and *accessible_attr*. We use Query 4 to get the generalization level of a particular attributes for (role, purpose).

Query 3. (retrieveData) SQL query for retrieving data from database.
SELECT $accessible_attr$
FROM $tables$

Query 4. (getLevel) SPARQL query for retrieving generalization level.
PREFIX Fi: <http://www.owl-ontologies.com/Finance2#>
PREFIX db: <http://biostorm.stanford.edu/db_table_classes/
          DSN_jdbc.mysql.//localhost.3306/db#>
SELECT ?gen_level
WHERE { $accessible_attr$?gen_level $role_purpose$. }

Once we have got data of the generalization level, the Data Generalization module uses Query 5 to get the generalization hierarchy for the microdata. Finally, the results are displayed to the user.

Query 5. (getGenData) SPARQL query for retrieving generalized data.
PREFIX rdfs: <http://www.w3.org/2000/01/rdf-schema#>
PREFIX Fi: <http://www.owl-ontologies.com/Finance2#>

SELECT ?gen_data
WHERE { $microdata rdfs:subClassOf ?gen_data.
          ?gen_data Fi:isLevel $gen_level. }

6   Privacy Preservation in Service Composition

In the previous sections we have discussed privacy preservation when request individual data services and proposed using ontology to provide different levels of data access. Ontologies are used to define roles and purposes and data generalization. Web service compositions are often needed to fulfill complicated service tasks.

Service consumers send a service request to a mediator, which can access a service repository, provided by service providers, and generate composite services to fulfill the given service request. Service discovery is an important step before services can be used or composed to create new services. The literature reveals several approaches on service discovery. Due to the lack of semantic understanding of syntactic approaches, which match service requests and descriptions of services based on keywords. Semantic based approaches are proposed to automatically generate service composition using ontologies. In our architecture framework, a mediator is not only responsible for semantic service discover but also responsible for ensuring privacy conservation when providing services to users, both atomic services and composite services.

A *web service S* is described with a functional description of input types I and output types O telling what service operation will do, a categorical description by inter-related keywords telling that the service operation does by using common terminology T of application area, a quality of service (QoS) description of non-functional properties such as response time, cost, availability, etc.

The QoS description is not needed for service discovery but is useful to select among alternative service compositions. In order to locate service it is critical to provide an adequate description, which will allow a search engine to discover the required services. For categorical description, the terminology has to be specified. This defines an ontology, for which we need to provide definitions of "concept" and relationship between them. In our previous work [24] we have defined terminology for service ontology using description logics.

In [24] one particular description logic, *DL-Lite* family (see [28]), is employed to define TBox. A *terminology* (or TBox) is a finite set T of assertions of the form $C_1 \sqsubseteq C_2$ with concepts C_1 and C_2 as defined by the grammar in [24]. Each assertion $C_1 \sqsubseteq C_2$ in a terminology T is called a *subsumption axiom*. As usual, we use the shortcut $C_1 \equiv C_2$ instead of $C_1 \sqsubseteq C_2 \sqsubseteq C_1$. For concepts, $\bot$ is a shortcut for $\neg \top$, and $\leq m.R$ is a shortcut for $\neg \geq m + 1.R$.

A *service repository* $\mathcal{D}$ consists a finite collection of services $s_i, i \in \{1, n\}$ together with a service terminology T. A service request R is a tuple (I_R, O_R), where I_R is the input concept that that the user can provide, O_R is the output concept the user expect. A service request can be represented as two special services, a starting service $s_0 = (\emptyset, I_R)$ and an end service $s_e = \{O_R, \emptyset\}$.

A service can be composed by a set of process expressions. The set of *process expressions* of a service composition is the smallest set $\mathcal{P}$ containing all elementary service constructs that is closed under sequential composition construct $\cdot$, parallel construct $\|$, choice $+$ and iteration $*$. That is whenever $p.q \in \mathcal{P}$ hold, then all $p.q$, $p\|q$, $p + q$ and $p*$ are process expression in $\mathcal{P}$, with p and q are component services. A service composition is a process expression with component services s_i, with each service associated with an input and output type I_i and O_i.

When we compose two services using sequential expression we need to check if one service match the other. A service s_j *fully matches* another service s_i if and only if the output type O_i and input type I_j are equivalent concepts or O_i subsumes concept I_j, i.e., $\mathcal{T} \models O_i \sqsubseteq I_j$. A service s_j *partially matches* another service s_i if $\mathcal{T} \models O_i \sqcap I_j \sqsubseteq \bot$.

Service composition plan can be naturally represented as Directed Acyclic Graphs (DAGs) [29]. Given a service repository $\mathcal{D}$ and a service request R, the service composition problem is to find a directed acyclic graph $G = \{V, E\}$ where V is the set of vertices and E is the set of edges of the graph, with one starting service s_0 and end service s_e and a set of intermediate vertices $\{V_1, \ldots, V_m\}$, which representing component services selected from the service repository. A service composition $\mathcal{S}$ is a *feasible* solution for the service request R if the following conditions are satisfied:

- All the inputs needed by the composition can be provided by the service request, i.e., $I_R \sqsubseteq I_\mathcal{S}$;
- All the required outputs can be provided by the composition, i.e., $O_\mathcal{S} \sqsubseteq O_R$;
- For each service s_j its inputs I_j should be provided by its proceeding services.

A service composition solution is *correct* if it is a feasible solution that satisfies the above mentioned criteria. There are many automatic service composition approaches proposed in the literature that can be used to generate service composition plans [30].

6.1   Privacy Aware Service Composition Framework

To preserve data privacy during service composition, data should be displayed at a level allowed for the user with a role and purpose. For this we define an architecture of accessing data via composite services, see Fig. 7. The Privacy-aware Executor filters the data to ensure that the usage rules are respected in order to preserve the privacy of the customer data. The Service Composition Generator uses existing approaches on automatic service composition. If it is QoS-aware service composition, then objective function of these approaches are optimizing quality of services.

6.2   Service Composition Generator

The role of Service Composition Generator is to design service composition plan that fulfill a service request from service consumers. A service consumer specifies

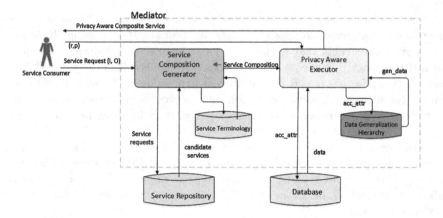

Fig. 7. Privacy aware service composition framework

functional requirement (I_R, O_R). The service consumer has a role and purpose of using the functionality. We can use the approach proposed in [29] generate service composite plan for a given composition task. A GraphEvol candidate may look like the representation shown in Fig. 8, with each block encoded as a graph node, and edges flowing from the input (start) towards the output (end) nodes. Note nodes with multiple children and/or multiple parents can also be included when necessary. Algorithm 1 describes the general procedure followed by GraphEvol.

Algorithm 1. Service Plan Generator: GraphEvol [29].

---

1. Initialise the population using a graph building algorithm.
2. Evaluate the fitness of the initialised population.
while *max. generations not met* **do**
    3. Select the fittest graph candidates for reproduction.
    4. Perform mutation and crossover on the selected candidates, generating offspring.
    5. Evaluate the fitness of the new graph individuals.
    6. Replace the lowest-fitness individuals in the population with the new graph individuals.

GraphEvol initializes a population of candidates that are encoded using non-linear data structures, evolves this population using crossover and mutation operators, and evaluates the quality of each candidate based on the nodes included in its structure. Figure 8 shows a graph representation of service composition. Mutation and crossover operators have been proposed to ensure the correctness of service composition. Based on users QoS requirement, a fitness function can formulated to evaluate the goodness of a service composition solution. Another important aspect of GraphEvol is that it uses a graph-building algorithm for creating new solutions, and for performing mutation and crossover. The general procedure for GraphEvol is shown in Algorithm 1. For a detailed discussion of graph mutation and crossover operations, the reader is referred to [29]. Once a service composition is generated a further data generation step is need to ensure data privacy to be preserved by the composite service.

6.3   Privacy Aware Service Executor

For a given service request, once a service composition plan is generated, the Privacy Aware Executor will execute the service composition plan and display data according to the role and purpose of the service requester. It takes the input, the role and purpose pair (r, p), and retrieves generalized data of the output O_R based on generalization hierarchies defined in previous sections.

The output O_S of privacy aware service composition is defined by (r, p). Data will be displayed according to the granularity of data defined by data generalization hierarchies $\mathcal{H}$. For each attribute in the output, a generalized value is presented instead of the original data retrieved from the database.

Algorithm 2. Privacy Aware Executor

Input: A composition plan represented as G, role-purpose (r, p), a set of data generalization hierarchies $\mathcal{H}$

Output: A privacy aware service composition execution S

foreach $o_i \in O_R$ **do**

　　If $level_i$ of o_i is higher than $Level\_1$

　　Retrieve generalized data of o_i from $H_i \in \mathcal{H}$

　　Else Retrieve data of o_i from database

　　Endif

During the process of service execution some sensitive data may be exchanged among component services. These may raise some threads. The challenge is to secure data exchange and release only data to avoid data attack in the middle of service composition. One way to do it is to apply data generalization for each component service before they are composed. However, the drawback of this approach is the computation overhead generated by replacing each component service with privacy aware component services. Therefore, our approach only retrieve generalized data for attributes in the output defined by the service request.

6.4   Example of Using the Privacy Aware Service Composition Framework

To illustrate the above mentioned framework we consider a service request from e-heath domain, in which a patient need to make an appointment with a nearby insurance officer for the purpose of buying an insurance. The service request R is defined by $I_R = \{Firstname, Lastname, Illness, Postcode\}$ and $O_R = \{Appointment(Firstname, Lastname, OfficerName, Illness, Location, Time)\}$. Assume the following is a service composition discovered by the Service Composition Generator.

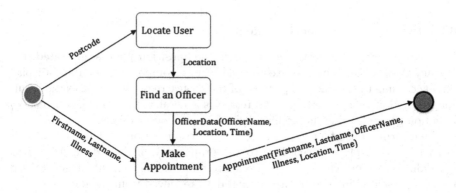

Fig. 8. An example of service composition represented as DAG

The output of the service composition contains an appointment that includes $\{Firstname, Lastname, Name, Location, Time, Illness\}$, where only generalized illness needs to be included in the appointment. The generated service composition plan will then be processed by the service executor to present generalized disease. As defined in Table 2 a disease data at generalization $Level\_2$ is presented to insurance officer.

7   Evaluation

This sections shows our evaluation of our proposed access model. First we demonstrate our proposed access model with prototypes of web services. Then we show performance of our proposed access model, comparing with two other access models.

7.1   Prototype Web Services

To demonstrate our web services architecture and the access model, we implement a web-based prototype system to provide web services, each of which provides data of different generalization levels based on roles and purposes to satisfy data privacy requirement in Sect. 2. We first set up our ontology in the Protégé ontology editor and knowledge acquisition system. Then we imported all attributes in the databases into Protégé via DataMaster plug-in, and assign access levels according to our ontology. Figure 9 shows three web services provided for LifeInsurance purpose to three roles: Admin, Insurance, and Bank.

Data released by web services satisfies usage rules in Table 2. We can see that for the same purpose *LifeInsurance* our system provide different web services to different roles. For example, the web service for insurance role presents generalized data of salary, deposit and illness. Our prototype system demonstrate that data privacy can be preserved while necessary information is released to users to meet their access purposes.

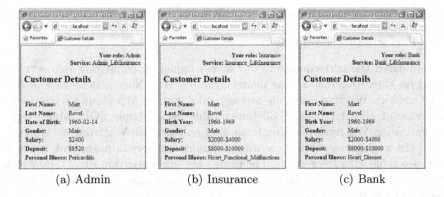

(a) Admin   (b) Insurance   (c) Bank

Fig. 9. The results accessed with LifeInsurance purpose by different roles

In the mean time for each role our system provides different web services for different purposes as shown in Fig. 10. We can see that the information provided to the same users are different, based on the purposes. The system reveals the information as least as possible to complete the purposes.

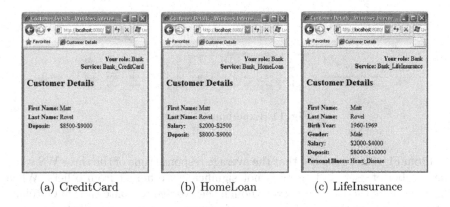

(a) CreditCard   (b) HomeLoan   (c) LifeInsurance

Fig. 10. The results from using different web services by Bank role

From above prototype web services we can see that our system can help preserve privacy of data because customer's information is only revealed to users with roles just for completing their purposes. Our system also supports generalization of data which overcomes the binary access decision (allowed or denied). The generalized data can help protect customer's privacy in some stages that the actual data does not have to be revealed. For example, the users can just use the approximate deposit amount or age range to complete their analysis tasks. Moreover, using ontologies for storing generalization hierarchies is more efficient than the approaches in [11–13] and therefore can provide better system performance.

7.2   Performance Evaluation

We evaluated the performance of our access model by comparing web services based on three different access models: Simple WS, None-generalization WS, and Ontology WS that uses our access model. Simple WS simply retrieves all the value of all attributes of Customer table using simple SQL, while None-generalization WS retrieves actual values of only accessible attributes. We evaluated the average response time with different numbers of users that access a database table with 100 records. The results are summarized in Fig. 11, on which the x-axis indicates the number of users while the y-axis indicates the average response time in milliseconds.

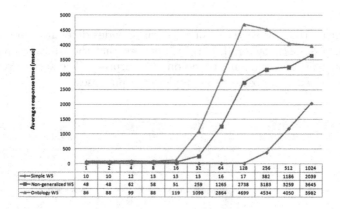

	1	2	4	8	16	32	64	128	256	512	1024
Simple WS	10	10	12	13	13	13	16	17	382	1186	2039
Non-generalized WS	48	48	62	58	51	259	1265	2738	3183	3259	3645
Ontology WS	86	88	99	88	119	1098	2864	4699	4534	4050	3982

Fig. 11. Experimental results.

From Fig. 11, we can see that the average response time of the three WS when the number of users is 1 to 16 is not significantly different from others. When there are more than 16 users, even though the average response time of Ontology WS is little longer than the other two web services, but it provides better privacy protection than the Simple WS, and more flexible access of data than None-generalization WS, with which access of data is limited to only allowed and denied.

7.3 Discussion

In this paper, we have proposed a privacy preservation approach based on roles and purpose of users. Compared to the approach in [3] that makes use of Hippocratic databases, our approach considers roles and supports data generalization. [20] proposes a role-based purpose-involved access control model to ensure privacy preservation, which needs to compute a set of conditional purposes implied by intended purpose using a propose tree. In our approach, we do not need to compute conditional purposes. Instead, we select suitable purposes by querying role-purpose ontology hierarchies using ontology query languages. Further, our purposed access model can provide a much finer-grained access control using data generalization ontology hierarchies that the approaches in [18,19]. In [8], the authors present an access control system for web services that combines trust-based decision policy and ongoing access control policy to create a secure protection system. The limitation of this approach is that the storage overhead increases dramatically when the number of records grows and the number of accessed attributes increases. Our approach has less storage overhead because we use ontology generalization hierarchy and therefore does not need to include all possible generalized data for each level.

8 Conclusion and Future Work

This paper proposes an access model to ensure privacy of data while sharing data via web services. The proposed access model employs ontologies to define access rules based on roles and purposes. The access model controls the information provided by web services to users to ensure privacy of customers can be preserved while tasks of service users can be accomplished. Our proposed access model can be used to define web services that release only information necessary for accomplishing a purpose. For service requests when a single existing service cannot fulfill the functional requirement we have discussed how service composition can be provide the required functionality, and presented a privacy-preserving framework for service composition. A prototype system has been implemented to demonstrate that different web services can be offered to service users based on their role and the purpose of their data access. Further, we have conducted an experimental evaluation to evaluate the performance of the access model. The results shows that our proposed access model can protect sensitive data of customers while sharing data via web service with reasonable and acceptable performance. Future work can be done to evaluate and analyze the storage overhead of our proposed approach. Also, we will extend our approach to support data owners to customize data accesses for each data user.

References

1. Ghani, N.A., Sidek, Z.M.: Privacy-preserving in web services using hippocratic database. In: International Symposium on Information Technology, vol. 1, pp. 1–5 (2008)
2. Bertino, E., Byun, J.-W., Li, N.: Privacy-preserving database systems. In: Aldini, A., Gorrieri, R., Martinelli, F. (eds.) FOSAD 2004-2005. LNCS, vol. 3655, pp. 178–206. Springer, Heidelberg (2005). doi:10.1007/11554578_6
3. Agrawal, R., Kiernan, J., Srikant, R., Xu, Y.: Hippocratic databases. In: 28th International Conference on Very Large Data Bases (VLDB), pp. 143–154 (2002)
4. LeFevre, K., Agrawal, R., Ercegovac, V., Ramakrishnan, R., Xu, Y., DeWitt, D.: Limiting disclosure in hippocratic databases. In: 30th International Conference on Very Large Data Bases (VLDB), pp. 108–119 (2004)
5. Agrawal, R., Kini, A., LeFevre, K., Wang, A., Xu, Y., Zhou, D.: Managing healthcare data hippocratically. In: ACM SIGMOD International Conference on Management of Data, pp. 947–948 (2004)
6. Massacci, F., Mylopoulos, J., Zannone, N.: Hierarchical hippocratic databases with minimal disclosure for virtual organizations. VLDB J. **15**, 370–387 (2006)
7. Laura-Silva, Y., Aref, W.: Realizing privacy-preserving features in hippocratic databases. In: IEEE 23rd International Conference on Data Engineering Workshop, pp. 198–206 (2007)
8. Li, M., Sun, X., Wang, H., Zhang, Y., Zhang, J.: Privacy-aware access control with trust management in web service. World Wide Web **14**, 407–430 (2011)
9. Xiao, X., Tao, Y.: Personalized privacy preservation. In: ACM SIGMOD International Conference on Management of Data, pp. 229–240 (2006)
10. Samarati, P., Sweeney, L.: Generalizing data to provide anonymity when disclosing information. In: ACM SIGACT SIGMOD SIGART Symposium on Principles of Database Systems, vol. 17, p. 188 (1998)
11. Sweeney, L.: Achieving k-anonymity privacy protection using generalization and suppression. Int. J. Uncertainty Fuzziness Knowl. Based Syst. **10**(05), 571–588 (2002)
12. Miller, J., Campan, A., Truta, T.M.: Constrained k-anonymity: privacy with generalization boundaries. In: Practical Privacy-Preserving Data Mining, p. 30 (2008)
13. Kisilevich, S., Rokach, L., Elovici, Y., Shapira, B.: Efficient multidimensional suppression for k-anonymity. IEEE Trans. Knowl. Data Eng. **22**, 334–347 (2010)
14. Omran, E., Bokma, A., Abu-Almaati, S.: A k-anonymity based semantic model for protecting personal information and privacy. In: IEEE International Advance Computing Conference, pp. 1443–1447 (2009)
15. Martínez, S., Sánchez, D., Valls, A., Batet, M.: The role of ontologies in the anonymization of textual variables. In: 13th International Conference of the Catalan Association for Artificial Intelligence, vol. 220, p. 153 (2010)
16. Domingo-Ferrer, J., Torra, V.: Disclosure control methods and information loss for microdata. In: Confidentiality, Disclosure, and Data Access: Theory and Practical Applications for Statistical Agencies, pp. 93–112 (2001)
17. Iqbal, Z., Noll, J., Alam, S., Chowdhury, M.M.: Toward user-centric privacy-aware user profile ontology for future services. In: 3rd International Conference on Communication Theory, Reliability, and Quality of Service, pp. 249–254 (2010)
18. Finin, T., Joshi, A., Kagal, L., Niu, J., Sandhu, R., Winsborough, W., Thuraisingham, B.: ROWLBAC: representing role based access control in owl. In: 13th ACM Symposium on Access Control Models and Technologies, pp. 73–82 (2008)

19. Cirio, L., Cruz, I.F., Tamassia, R.: A role and attribute based access control system using semantic web technologies. In: On the Move to Meaningful Internet Systems Workshops, pp. 1256–1266 (2007)
20. Kabir, M.E., Wang, H., Bertino, E.: A role-involved purpose-based access control model. Inf. Syst. Frontiers, 1–14 (2012)
21. Tumer, A., Dogac, A., Toroslu, I.H.: A semantic based privacy framework for web services. In: Proceedings of ESSW (2003)
22. Gruber, T., et al.: A translation approach to portable ontology specifications. Knowl. Acquisition **5**, 199–220 (1993)
23. Wang, Y., Liu, W., Bell, D.: A concept hierarchy based ontology mapping approach. In: Bi, Y., Williams, M.-A. (eds.) KSEM 2010, pp. 101–113. Springer, Heidelberg (2010)
24. Ma, H., Schewe, K.D., Wang, Q.: An abastract model for service provision, search and composition. In: Proceedings of 2009 IEEE Asia-Pacific Services Computing Conference (APSCC), pp. 95–102. IEEE (2009)
25. Li, M., Wang, H., Plank, A.: Privacy-aware access control with generalization boundaries. In: 32nd Australasian Conference on Computer Science, pp. 105–112 (2009)
26. Talouki, M., NematBakhsh, M.a., Baraani, A.: K-anonymity privacy protection using ontology. In: 14th International CSI Computer Conference, pp. 682–685 (2009)
27. Iyengar, V.S.: Transforming data to satisfy privacy constraints. In: 8th ACM SIGKDD International Conference on Knowledge Discovery and Data Mining, pp. 279–288 (2002)
28. Baader, F., et al. (eds.): The Description Logic Handbook: Theory, Implementation and Applications. Cambridge University Press, New York (2003)
29. da Silva, A., Ma, H., Zhang, M.: GraphEvol: a graph evolution technique for web service composition. In: Chen, Q., Hameurlain, A., Toumani, F., Wagner, R., Decker, H. (eds.) DEXA 2015. LNCS, vol. 9262, pp. 134–142. Springer, Heidelberg (2015)
30. da Silva, A.S., Ma, H., Zhang, M.: A graph-based particle swarm optimisation approach to QOS-aware web service composition and selection. In: 2014 IEEE Congress on Evolutionary Computation (CEC), pp. 3127–3134. IEEE (2014)

Author Index

Printed in the United States
By Bookmasters